Thrive in Bioscience | Revision Guides

Other titles in the Thrive in Bioscience series

Thrive in Biochemistry and Molecular Biology

Lynne S. Cox, David A. Harris, and Catherine J. Pears

Thrive in Cell Biology

Qiuyu Wang, Chris Smith, and Emma Davis

Thrive in Ecology and Evolution

Alan Beeby and Ralph Beeby

Forthcoming: **Thrive in** Human Physiology

Ian Kay and Gethin Evans

Thrive in Immunology

Anne Cunningham

Thrive in Genetics

Alison Thomas
Senior Lecturer in Genetics, Anglia Ruskin University

Thrive in Bioscience | Revision Guides

OXFORD
UNIVERSITY PRESS

Great Clarendon Street, Oxford, OX2 6DP,
United Kingdom

Oxford University Press is a department of the University of Oxford.
It furthers the University's objective of excellence in research, scholarship,
and education by publishing worldwide. Oxford is a registered trade mark of
Oxford University Press in the UK and in certain other countries

British Library Cataloguing in Publication Data
Data available

ISBN 978-0-19-969462-4

Printed and bound in Great Britain by
CPI Group (UK) Ltd, Croydon, CR0 4YY

Contents

Contents

Contents

Four steps to exam success

1 Review the facts

- This book is designed to help you learn quickly and effectively:
  - information is set out in bullet points, making it easy to digest,
  - clear, uncluttered illustrations illuminate what is said in the text,
  - key concept panels indicate the essential learning points for a topic.

2 Check your understanding

- Try the questions in each chapter and online multiple-choice questions to reinforce your learning.
- Download the flashcard app to master the essential terms and phrases.

3 Take note of extra advice

- Look out for hints for getting those precious extra marks in exams.

4 Go the extra mile

- Explore other sources of information – including genetics textbooks – to take your knowledge and understanding one step further.

Go to the Online Resource Centre for more resources to support your learning, including:

- online quizzes, with feedback,
- a flashcard glossary, to help you master the essential terminology.

online resource centre

www.oxfordtextbooks.co.uk/orc/thrive/

Using this guide

This book is designed to be an aide memoire and a quick check-up source rather than an authoritative text. You will need to have revised material from your lecture notes, textbooks, and recommended reading lists in order to get full value from this book. It aims to include core material that is central to an understanding of genetic processes. However, courses differ, so not all the material here may be covered in your course, and there may be bits we haven't included here that you do need to know about. It is also worth checking out the other revision guides in this series, for example those on biochemistry and molecular biology, and cell biology, according to your own course requirements.

Exam technique

When answering **short questions**, ensure that your answer is concise and precise, containing all the information required, but no waffle or irrelevant information. In some cases, it is okay to use bullet points, but not in others, so make sure you choose a format appropriate to your own university's exams. Although exam styles vary, examiners will always appreciate scientific accuracy, relevance, and logical argument.

There are multiple choice tests online linked to each chapter of the book, so you can check your progress in revision. As part of your revision use the glossary, which defines important terms.

 online resource centre www.oxfordtextbooks.co.uk/orc/thrive/

For **multiple-choice questions**, try to come up with the correct answer first, then check the answers given—be particularly careful when you have to identify correct combinations of answers. For example:

Which of the following is true about eukaryotic translation?
   i) The ribosome is made up of 40S and 60S subunits forming a 100S complex.
   ii) The ribosome is made up of 40S and 60S subunits forming an 80S complex.
   iii) The AUG start codon of the mRNA is initially located in the P site of the ribosome.
   iv) The AUG start codon of the mRNA is initially located in the A site of the ribosome.
   v) Peptide bond formation is mediated by an RNA enzyme (a ribozyme).

Answers:

A: i, iv, v.   **B:** i, iii, v.   **C:** ii, iii, v.   **D:** ii, iv, v.   **E:** ii, iv.

(the correct answer is C)

When tackling **exam essays**, remember the following.

Read the question carefully and try to identify exactly what the examiner is asking for (e.g. is the question limited to prokaryotes or eukaryotes, or should you include both? Are details of a single genetic concept or technique required, or do you need to integrate material from various parts of your course to answer the question fully? If so, do you know all those topics well enough to provide a balanced answer?).

For a 1-hour essay, spend at least 5 minutes planning at the start—it may seem scary when everyone else appears to have launched straight in, but your essay will be coherent and well structured, and planning makes sure that you don't forget to include critical points and that you can balance the material appropriately. Preparation is key here—it really helps if you have already planned answers to a number of different essay questions on each topic; the chances are high that a similar question will be asked in your exam. Remember, the pressure is much less if you are confident that you have prepared thoroughly. There is no single correct way to plan an essay—some people make mind maps, others prefer lists—but do make sure that you don't waste time writing full prose (save that for the essay; you can use abbreviations as much as you like in a plan). As you are writing, do keep checking your plan to make sure you include everything you think important.

Avoid waffle—every word needs to count so make sure it's scientific and conveys information quickly, clearly, and coherently. Do make sure you use technical terms correctly and spell scientific words carefully. Examiners are not just looking for factual content, but evidence that you understand the material; you can demonstrate this by using a clear essay structure (subheadings can really help guide the examiner in your thought processes—check if it's permitted at your institution) and by providing a cogent argument.

Try to include an introduction, the core material, and, if there's time, a rounded conclusion.

The **introduction** should contain a *definition* of the key terminology of the question at the start of your answer. You can pick up marks quickly this way and begin the essay in a convincing manner (e.g. 'DNA replication is the template-directed synthesis of a polymer of deoxyribonucleotides mediated by DNA polymerases involving formation of phosphodiester bonds…' sounds rather more scientific than 'DNA replication is important and happens in all cells every time they divide'—a statement that is equally true but does not display scientific knowledge). You can also set the scope of your essay in the introduction, for example, 'I shall discuss predominantly eukaryotic DNA synthesis, but will mention the process in prokaryotes where appropriate…'.

Paragraphs in the **main body** of the essay should contain the following.

- Key **concept**, with details of the genetic process (as relevant).
- Specific **named example**(s).
- **Experimental evidence** to support the ideas (even when not specifically asked for in the question). By providing support, you show the examiner that you

understand how information was obtained and how strong the evidence is behind the idea—it's a great way of bumping up marks. As you progress through your degree, you will be expected to refer directly to experiments in the primary research literature, so get into the habit of providing experimental evidence early on.

- If appropriate, do illustrate your answer with an **annotated diagram**, for example draw out the genetic process *simply* but clearly (remember, you are being marked on scientific content not artistic merit). Diagrams are only of exam value if they are fully annotated (i.e. they have descriptive labels) and **don't take longer to show ideas than it would have taken you to write text describing them**.

- In most universities, it is fine to use colour in diagrams, but do check first. For example, illustrating the complex processes of DNA recombination is made much simpler by using a different colour for each strand of DNA. The aim of this book was to produce simple diagrams that you might be able to reproduce easily under exam conditions (though they are in black and white for printing purposes).

- Don't repeat the same material in diagrams and text, but do introduce each diagram, e.g. *One elongation cycle during translation*. Remember to use full labels that are clear, legible, and informative.

- Ensure that you leave sufficient space around your diagrams so the reader can easily see what you are illustrating—don't squeeze them into the margins or wrap text around them.

- Comparative tables can also be really helpful to convey factual information rapidly and show the examiner that you can identify key concepts or stages in a process, and relate the details to those concepts (e.g. *Key differences between prokaryotic and eukaryotic translation*). However, ensure you don't rely wholly on such tables, as you will also be marked on the coherency of your discussion.

- Round off each paragraph with how the information you have just presented addresses the question.

Try to include a **conclusion** where you can argue for/against (especially if it's a 'Discuss…'-type essay). You can throw in the odd quirky example here if it didn't fit well into the rest of the essay, but make sure it's relevant. If there are controversies in the field, you can mention them here. Don't waste time repeating things you have already mentioned. You could even highlight what further knowledge is required to fully understand the process (but make sure it's a real gap in knowledge, not simply that you didn't know it!).

Good luck in your exams and do remember that genetics is a subject to enjoy—it's not just about passing exams!

Alison Thomas,

Cambridge,
October 2012

1 Chromosomes and Cellular Reproduction

One or more chromosomes contain the genetic material of a cell. Mitosis and meiosis distribute chromosomes to daughter cells.

Key concepts

- Living organisms are classified as prokaryotes or eukaryotes.
- Eukaryotic cells have a more complex structure.
- Each eukaryotic species has a characteristic number of chromosomes.
- Different chromosomes of a species are distinguished by size and position of the centromere.
- Each eukaryotic chromosome contains a single long linear DNA molecule which is tightly packed.
- The sex of an individual is often determined by a pair of sex chromosomes.
- The **cell cycle** describes the different stages that a eukaryotic cell passes through from one cell division to the next.
- Mitosis results in the production of two genetically identical cells.
- Meiosis leads to the formation of four genetically variable cells with half the chromosome number.

1.1 CELLS

Living organisms are classified into two major groups, the **prokaryotes** and **eukaryotes**, which differ in their cellular structure and organization of the genetic material.

Prokaryotes

- Prokaryotes are unicellular organisms with a simple cellular structure.
- They lack nuclei and other membrane bound organelles.
- Their genetic material is in the cytoplasm in the **nucleoid** region; typically as one small circular **chromosome**.
- There are two distinct groups of prokaryotes:
  - **Eubacteria** (true bacteria), which include most of the bacteria
  - **Archea** (ancient bacteria), a diverse group generally found in extreme habitats, such as hot springs and ocean depths.

Eukaryotes

- Eukaryotic organisms include animals, plants, fungi, and protists.
- The cell is compartmentalized into membrane-bound organelles.
- Most of their genetic material is found in the **nucleus**, where it is partitioned into separate chromosomes.
- A small amount of their DNA is found in organelles: mitochondria and chloroplasts. For example, 38 of the estimated 23,500 human genes are encoded by mitochondrial DNA.

1.2 EUKARYOTIC CHROMOSOMES

- A eukaryotic chromosome consists of DNA and associated proteins.
- It is a self-replicating structure that carries in the DNA nucleotide sequence a linear array of genes.
- A cell's DNA requires highly organized packaging to fit into a nucleus. The resulting chromosomes have distinctive features.

Chromosome number

- There is great variability in chromosome number between species (Table 1.1).
- The number of chromosomes is constant in each cell of a given species.
- **Somatic cells** (any cell other than sex cells) contain two copies of each chromosome.

Species name	Chromosome number	Species name	Chromosome number
Elephant *Loxodonta africana*	56	Garden pea *Pisum sativum*	14
Opossum *Didelphis virginiana*	22	Adder tongue fern* *Ophioglossum vulgatum*	1200 or 1260
Silkworm *Bombyx mori*	56	Potato *Solanum tuberosum*	48
Human *Homo sapiens*	46	Pineapple *Ananas comosus*	50
Goldfish *Carssinus auratus*	94	Cultivated wheat *Triticum aestivum*	42
Alligator *Alligator missispiensis*	32	Redwood *Sequoiadendron giganteum*	22
Dog *Canis familiaris*	78	Rice *Oryza sativa*	24

Table 1.1 Diploid chromosome number in a selection of animal and plant species
*This fern has the highest known chromosome number of any living species.

- One set of chromosomes is inherited from the female parent (termed the **maternal** chromosomes). The other set (termed **paternal** chromosomes) is inherited from the male parent.
- A matching chromosome pair (i.e. a maternal and paternal chromosome) is termed **homologous** chromosomes, or **homologues**.
- Homologues are similar in size, physical structure, and gene composition and order.
- Homologous pairs of chromosomes are also termed **autosomes**.
- Cells with two sets of chromosomes are **diploid** or **2n**. The suffix '-ploid' refers to sets of chromosomes.
- Gametes (sex cells), containing one copy of each chromosome, are **haploid**.
- The haploid chromosome set (**n**) is also termed the **genome**: one complete set of a species' genetic instructions.

Chromosome morphology: gross structure

- Chromosomes can only be seen as discrete structures during nuclear division (mitosis and meiosis), when the chromosomes become highly condensed.
- Chromosomes are generally seen as double (replicated) structures held together by a centromere (Figure 1.1).
- Key chromosomal features are:
  ○ the relative lengths of a cell's different chromosomes
  ○ the **centromere** that divides a chromosomes into the upper or **p arm**, and the lower or **q arm**. Spindle fibres attach to centromeres during nuclear division.
- Depending upon the position of the centromere, chromosomes are described as:
  ○ **metacentric**—the p and q arms are approximately the same length
  ○ **acrocentric**—the p arm is shorter than the q arm
  ○ **telocentric**—the centromere is very near one end of the chromosome.

Eukaryotic chromosomes

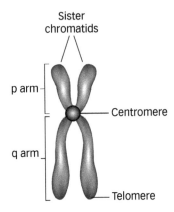

Figure 1.1 A eukaryotic chromosome: key features.

- Each chromosome replica is known as a **chromatid**.
- The two chromatids of a replicated chromosome are known as **sister chromatids**.
- Additional constrictions or expanded regions (**knobs**) are sometimes present.
- **Telomeres** stabilize the ends of chromosomes.
- They are specialized terminal DNA sequences that ensure the maintenance and accurate **replication** of the two ends of a linear chromosome.
- Distinctive different chromosome morphologies arise occasionally, as with **polytene** and **lampbrush chromosomes** (Table 1.2).
- A **karyotype** is a visual display of a cell's chromosomes. Homologous chromosome pairs are arranged in descending size order (Figure 1.2).

Looking for extra marks?

Each time a cell divides telomeres shorten slightly. Once the telomeres are reduced to a certain critical length, a cell can no longer divide and becomes inactive or 'senescent'. Signals are transmitted that initiate apoptosis (programmed cell death).

Preparation of a karyotype

- Karyotypes are prepared from actively dividing cells, e.g. white blood cells, bone marrow cells or plant meristematic cells.
  - Cells are cultured for 48–72 h.
  - **Colchicine**, an inhibitor of spindle microtubule assembly, is added to the culture to arrest the cells in metaphase of mitosis.
  - Harvested cells are placed on a slide in a hypotonic solution, which ruptures the nuclei.
  - A chromosomal stain is added.
- Depending on the stain used, different chromosomes show characteristic light and dark **banding patterns**. These patterns are useful in identifying individual chromosomes and in detecting structural changes.

Chromatin	Description
Euchromatin	• Loosely packed, actively transcribed DNA
Heterochromatin	• **Constitutive:** heterochromatin in a permanent compacted state—found at centromeres and telomeres • **Facultative:** compaction of this chromatin varies according to cell type and stage of development, i.e. the need for a gene product
Nucleolar organizing regions (NORs)	• The chromosome segments containing the tandemly repeated **rRNA** genes loop out from the chromatin fibre. Complexed with proteins involved in rRNA synthesis and processing, these regions are visible in light microscopes as **nucleoli**
Barr body	• An inactivated **X chromosome**; present as facultative heterochromatin

Chromosomes	Description
Polytene	• Giant interphase chromosomes formed from repeated rounds of DNA replication without cell division. The DNA copies (can be several thousand) align parallel • Giemsa staining produces light and dark bands visible under light microscope • Actively transcribed genes visible as **chromosomal puffs:** informative for gene expression studies • Found in salivary gland cells of Diptera larvae
Lampbrush	• Brush-like chromosome; results from decondensation of multiple transcribing genes • Transitory structures formed during prophase I of meiosis in oocytes of most animals (not mammals)
Artificial	• Made to clone large fragments of DNA (Chapter 12) • Contain a centromere, telomeres, an origin of DNA replication and the cloned DNA fragment • Yeast, mammalian, and bacterial artificial chromosome (YACs, MACs and BACs) are all used routinely

Table 1.2 Different types of chromatin and chromosomes

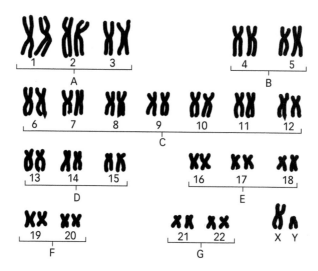

Figure 1.2 Karyotype of the chromosomes of a normal male human. Each chromosome is presented as a pair of chromatids. Autosomes are placed in six groups (A–G) according to size.

Eukaryotic chromosomes

- Good quality chromosomal spreads, i.e. those with a minimum number of overlapping chromosomes, are photographed.
- Individual chromosomes are cut out and arranged as a karyotype. This process is generally done digitally.
- The most common chromosome stain is **Giemsa**. It binds to A–T rich regions of DNA, producing **G-banding**. Dark bands indicate highly condensed **heterochromatin**, while light bands contain less condensed **euchromatin** (Table 1.2).
- Different uses of Giemsa dye produce R-banded and C-banded chromosomes.
- Staining chromosomes with quinacrine produces Q-banded chromosomes.
- An **ideogram** is a black and white schematic representation of banded chromosomes.

Looking for extra marks?

Karyotypes are used to predict human genetic disorders. They identify chromosome **mutations**, for example missing or extra chromosomes, or a chromosomal rearrangement.

Detailed chromosome structure

- Tremendous folding and packaging of DNA occurs within a eukaryotic chromosome.
- The average packing ratio of DNA (i.e. length of DNA to length of chromosome) is 10,000:1.
- The degree of DNA packaging
  - changes at different stages of the cell cycle: DNA is less tightly packed in an **interphase** chromosome compared with a mitotic one
  - depends on the transcriptional status of a gene: actively transcribed genes are more loosely packed than inactive ones.
- DNA packaging follows a set hierarchical organization (see Figure 1.3).

Nucleosomes

- The **nucleosome** core consists of a **histone** octamer, comprising two molecules of each of histones H2A, H2B, H3, and H4.
- 1.7 turns of DNA, or 145–147 nucleotide pairs, wrap around the histone core.
- A fifth histone, H1, acts as a clamp. It binds to the nucleosome where the DNA enters and leaves, holding the DNA in place. H1 binds a further 20–22 nucleotides (Figure 1.4).
- All five histones have a high percentage of the positively charged **amino acids** arginine and lysine. Thus, histones have a positive charge, which attracts the negatively charged DNA.

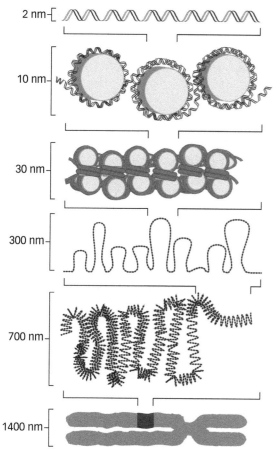

The first level of packaging, the 10nm chromatin fibre, is achieved by the DNA double helix winding around histone cores forming nucleosomes: the 'beads-on-a-string' structure. This is the form of actively transcribed DNA and is also termed euchromatin.

The second level involves the nucleosomes coiling and compacting into a 30nm fibre or solenoid. This is the form of heterochromatin in which genes are inactive (Table 1.4).

Higher level packaging occurs during mitosis and meiosis when the 30nm fibre loops around a protein scaffold, and condenses to produce a 'chromosome'.

Figure 1.3 The different levels of DNA packaging into a chromosome.

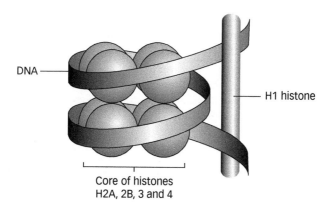

Figure 1.4 Structure of a nucleosome.

- Most of the arginine and lysine residues are in the histones' flexible tails that extend from the nucleosome core and closely contact the DNA.
- Chemical modification of the histone tails is key to changes in chromatin structure during DNA replication and **transcription**. Generally:
  - methylation inactivates genes
  - acetylation switches genes on.
- There is joining or **linker DNA**, generally 30–40 nucleotides, between nucleosomes.
- Thus, the nucleosome periodicity is 1 per 200 nucleotides.
- The amino acid sequence of histones, and so nucleotide sequence of histone genes, is highly conserved across phyla.
- A variety of non-histone proteins are also associated with DNA, e.g. RNA polymerases and regulatory proteins.

> ### Check your understanding
>
> 1.1 List structures that (a) prokaryotic and eukaryotic cells have in common, (b) distinguish prokaryotic and eukaryotic cells.
> 1.2 How many chromatids would you expect to see if a karyotype was prepared from salivary gland cells of the bluebottle fly *Calliphora vomitoria*; 2n=12.
> 1.3 What would be the consequences of neutralizing the positive charge of histone proteins?

1.3 SEX CHROMOSOMES AND SEX DETERMINATION

- In many animal and plant species the sex of an individual is determined by a specific pair of chromosomes: the **sex chromosomes**.
- Generally there are two different sex chromosomes. The presence of
  - two identical sex chromosomes determines one sex (XX or ZZ chromosomes)
  - two different chromosomes determines the other sex (XY or ZW chromosomes).
- In some species it is the number of a single sex chromosome (XX or XO, where O denotes the lack of a second sex chromosome) that determines the sex of an individual (Table 1.3). The critical sex determining factor is the ratio of X chromosomes: autosome chromosome sets (A):
  - females when X/A=1.0
  - males when X/A=0.5.
- Occasionally, in a few insect species (e.g. bees, wasps, and ants), the number of chromosome sets (or ploidy) determines the sex of an individual:
  - unfertilized haploid eggs develop into males
  - fertilized diploid eggs produce females.

	Male	Female
Mammals, some insects, many dioecious angiosperms	XY	XX
Birds, butterflies and moths, some fish	ZZ	ZW
Some insects, *Caenorhabditis elegans*	XO*	XX
Bees, wasps, ants	Haploid	Diploid

Table 1.3 Chromosomal mechanisms of sex determination
*Zero denotes lack of a second sex chromosome.

- Each chromosome of a sex chromosome pair (XY or ZW) carries different genetic information.
- This is in contrast to all other pairs of eukaryotic chromosomes which are homologous and carry genetic information for the same set of inherited characteristics.
- The X and Y chromosomes, but not the Z and W chromosomes, differ morphologically.
- Individuals with identical sex chromosomes are termed the **homogametic sex**, i.e. they produce one type of gamete (with regard to sex chromosomes).
- Individuals who have two distinct sex chromosomes are the **heterogametic sex**, producing two types of gametes.
- Consider the XX/XY sex determining system:
  ○ all gametes produced by an XX female have an X chromosome
  ○ XY males produce 50% of gametes with an X, and 50% of gametes with a **Y chromosome.**
- This homogametic/heterogametic system ensures equal sex ratios (Figure 1.5).
- For the ZZ/ZW system, males (ZZ) are homogametic and females (ZW) are heterogametic.

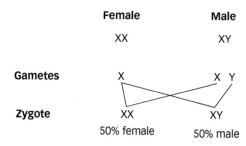

Figure 1.5 Mammalian sex determination.

Looking for extra marks?

A complex XY sex determination system occurs in monotremes. For example, the duck-billed platypus has ten sex chromosomes! Females are 10X and males are 5X5Y producing XXXXX and YYYYY sperm.

Sex chromosomes and sex determination

X and Y sex chromosomes

- X chromosomes are larger than Y chromosomes and contain many genes that encode non-sexual related traits.
- There are two small regions of homology, the **pseudoautosomal region,** between X and Y chromosomes (Figure 1.6). This region ensures X and Y chromosomes pair during prophase 1 of male meiosis and then segregate into separate cells.
- The X and Y pseudoautosomal regions behave like autosomal regions. Recombination occurs between these regions.
- To date, 24 genes have been identified in the pseudoautosomal region.

(i) SRY gene

- The presence or absence of a Y chromosome is the critical factor that determines the sex of an individual.
- There is a male-sex-determining gene on the Y chromosome. In humans this is the **SRY** (Sex-determining Region of the Y) gene.
- The SRY gene acts as a signal to set the development pathway towards maleness. It initiates a cascade of hormone changes.
- The 'default' developmental pathway in humans is female, which occurs in the absence of a functioning SRY gene.
- The importance of the SRY gene is emphasized by the following genetic situations:
  - a mutation in the SRY gene or SRY gene transfer from Y to X can result in XY females and XX males respectively
  - XXY individuals are male; XO are female.

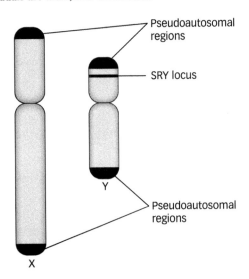

Figure 1.6 Human sex chromosomes.

(ii) X chromosome inactivation

- **X-inactivation** or **lyonization** ensures that only one of the two X chromosomes is functionally active in each female cell.
- This **dosage compensation** mechanism ensures that, as in male cells, functional product is produced from genes on just one X chromosome.
- Early in female embryological development, one of the two X chromosomes in each cell is randomly and permanently inactivated.
- Once an X chromosome is inactivated (designated **Xi**), it remains inactive throughout the lifetime of the cell and in all its derived cells.
- The inactivated X chromosome is highly condensed and visible as a **Barr body**: a densely staining, rice grain-shaped mass close to the nuclear membrane in interphase nuclei.
- Inactivation is initiated at a specific **X inactivation centre** (*Xic*) and spreads along the chromosome.
- **Methylation** (adding of CH_3 to CG dinucleotides) and histone deacetylation maintains inactivation.
- X inactivation is reversed in germ cells. Thus, it is an **epigenetic** phenomenon.

(iii) ZZ and ZW chromosomes

- Considerably more is known about the genetic nature of the X and Y chromosomes compared with the ZW chromosomes.
- The key sex determining factor is unknown in species using the ZW system.
- There is no homology between the ZW and XY chromosomes. Indeed, the greatest homology is between Z and mammalian chromosome 9.

Looking for extra marks?

Human manifesting female carriers of X-linked disorders result from skewed X inactivation, i.e. the majority of a woman's normal X chromosomes were inactivated embryologically.

Non-genetic sex determination

- In reptiles and various invertebrate species sex is determined by environmental and other non-genetic factors:
  - the temperature of the environment after fertilization
  - the physical location of the maturing individual
  - the age of the individual.
- In reptiles, the temperature of the eggs during a critical phase of development determines the sex of the developing individual.

	Male	*Female*
Alligators and lizards	Warm temperature	Cool temperature
Tortoises and turtles	Cool temperature	Warm temperature
Slipper limpet, *Crepidula fornicata*	Close to developing female	Distant to developing female
Anemone fish (*Amphiprion* spp.)	Young adults	Older adults

Table 1.4 Non-genetic mechanisms of sex determination

- Eggs exposed to high temperatures during this critical time produce males in some species, females in others. Intermediate temperatures can result in both males and females (Table 1.4).
- For example, the final phase of development decides the sex of the European pond turtle (*Emys obicularis*). Higher temperatures increase the activity of the enzyme aromatase, which converts testosterone into oestrogen, and so promotes female development.
- In a few marine invertebrate species proximity of a female to a maturing larva determines its sex. For example, larvae of the slipper limpet, *Crepidula fornicata*, develop into males if near a developing female and into females if further away from a female.

Check your understanding

1.4 How is sex determined in birds?

1.4 THE CELL CYCLE

- During its existence a somatic eukaryotic cell passes through a series of coordinated events termed the cell cycle (Figure 1.7).

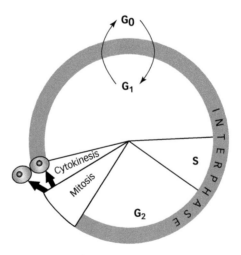

Figure 1.7 Key features of a cell cycle.

State	Phase	Details
Interphase	G$_1$ (Gap 1)	Cells grow and carry out their normal function. Duration highly variable. G$_1$ *checkpoint*: readiness for DNA synthesis
	S (DNA synthesis)	DNA replication occurs
	G$_2$ (Gap 2)	Cell growth and normal function continue. G$_2$ *checkpoint*: readiness for mitosis
Cell division	M (Mitosis)	Cell growth and function ceases. Orderly segregation of chromosomes. Cytokinesis follows mitosis *Metaphase checkpoint*: all chromosomes aligned at equator
Quiescent or senescent	G$_0$	Fully differentiated cells have left cycle, and stopped dividing

Table 1.5 Key events of the cell cycle

- The cell cycle has two main phases: interphase and mitosis.
- During interphase a cell grows, replicates its DNA and accumulates all the proteins needed for mitosis.
- Interphase commences with **G$_1$**, during which cells are carrying out their normal function.
- Preparations for division begin with cells entering the **S phase** when DNA is replicated, followed by the **G$_2$** phase from which cells enter mitosis.
- The length of G$_1$ is highly variable. Once cells pass a specific **R** or **restriction point** they are committed to completing a cycle. S, G$_2$ and M have relatively fixed lengths.
- In eukaryotes most cells do not divide regularly. They are removed from cycling and enter G$_0$, i.e. are *quiescent* and require stimulation before they will divide. Only specific tissues, such as bone marrow in animals and root tips in plants, have a high proportion of cycling cells.

Regulation of the cell cycle

- Progression through the cell cycle occurs in a defined sequence of events. It is monitored by a series of **checkpoints**, which cells cannot pass unless certain criteria are met.
- There are three major check points (Figure 1.8).
  - (i) **G$_1$/S boundary**—the integrity of the genome is checked. Damaged DNA is **repaired** before progression into S occurs. Replication of damaged DNA could result in mutations.
  - (ii) **G$_2$M boundary**—cells with uncompleted replication cannot enter M.
  - (iii) **Mid-M checkpoint**—mitosis cannot progress until the spindle is organized and all chromosomes are attached to spindle fibres.
- Two key groups of proteins control progression at these checkpoints: **cyclin dependent kinases (CDKs)** and **cyclins**.

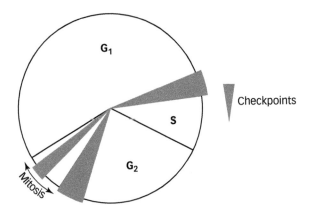

Figure 1.8 Key cell cycle checkpoints.

- CDKs trigger transition from G_1 to S, G_2 to M, and through M via phosphorylation of key proteins. These activate or inactivate, as appropriate.
- Cyclins indicate which proteins the CDKs phosphorylate. For example, one CDK-cyclin binds to the nuclear structural lamin proteins, initiating dissolution of the nuclear membrane at the onset of mitosis.
- Gene regulation turns on and off the synthesis of specific cyclins, and their rapid degradation once they have completed their task ensures their orderly appearance during the cell cycle.
- Many of the genes encoding the cyclins and CDKs are universal, e.g. the cyclin D–CDK4 complex initiates transfer from G_1 to S in all eukaryotes.

Looking for extra marks?

Genes involved in checkpoints are often mutant in tumours. For example, both copies of tumour suppressor p53 are mutated in ~40% of cancers. Normal p53 activates a pathway of DNA repair before entry into S.

1.5 MITOSIS

- Life continues from one generation to the next because chromosomes, cells, and organisms reproduce.
- Somatic cells reproduce through **mitosis** (nuclear division) followed by **cytokinesis** (division of the cell).
- Mitosis forms two cells, each with the same chromosome number and genetic complement as the parental cell (Table 1.6).
- Mitosis provides cells:
  - (i) for growth (zygote to mature adult)

Stage	Key events	
Prophase	• Chromosomes condense (shorten and thicken) • Each chromosome is present as a pair of sister chromatids attached at the centromere • The nucleolus disappears • RNA transcription virtually ceases **Centrioles** (seen in all eukaryotes except for higher plants) migrate to opposite ends (**poles**) of the cell • A star of microtubules radiates from the centrioles forming **asters** • A **spindle** begins to form outside the nucleus. It consists of many fibres, each an aggregate of microtubules made of the protein **tubulin** • The nuclear membrane breaks down	**Early prophase** Centrioles Chromosomes visible as single structures Nucleolus **Late prophase** Cytoplasm Plasma membrane
Metaphase	• Spindle fibres extend from opposite poles towards a central axis—the **equator** • Spindle fibres link with **kinetochores**, protein attachment sites on the centromere of each chromatid • At the poles spindle fibres link with centrioles • Each pair of chromatids is aligned at the equator • Completion of spindle formation is a crucial cell cycle checkpoint	**Metaphase** Spindle fibres
Anaphase	• Sister chromatids are now visible • Each chromatid with its attached kinetochore is pulled to opposite poles • Chromatid migration requires the activity of ATP-dependent **motor proteins** • At this point chromatids are once again termed chromosomes • Chromosome arms dangle behind centromeres during movement to poles	**Anaphase**
Telophase	• Chromosomes cluster at opposite poles • Chromosomes uncoil forming the long diffuse chromatin threads of interphase nuclei • Nuclear membranes/envelopes form around each chromosome group • Nucleolus reappears • Cytokinesis begins	**Telophase** **Interphase** Chromatin

Table 1.6 Details of each mitotic stage

(ii) repair of damaged tissues

(iii) cellular replacement, e.g. gut and skin.

- Mitosis begins with **prophase**, during which the nuclear envelope breaks down and the duplicated chromosomes condense.
- At **metaphase** the chromosomes align in the middle of the cell.
- During **anaphase** the chromatid pairs separate, with one copy of each pair moving to each end of the cell.
- In **telophase** each chromatid (once again termed a chromosome) reaches one of the two poles and nuclear envelopes reform around each chromosome set. Cytokinesis begins during telophase.
- Mitosis involves the coordinated activity of a variety of proteins that cause chromosomal condensation, produce spindles, dissolve, and reform the nuclear membrane and other mitotic functions.
- Details of each stage are given in Table 1.6.

Cytokinesis

- During cytokinesis partitioning of the cytoplasm occurs.
- The mechanism differs between animals and plants as plants need to additionally construct a new cell wall.
- In **animal** cells, the **cell furrow** develops in the cytoplasm between the two new nuclei. The cell membrane constricts the cell in a similar manner to a string loop being pulled around the middle of a balloon.
- In **plant** cells, vesicles containing wall material gather at the metaphase equator. These fuse to produce the **cell plate** against which new wall is laid down.

1.6 MEIOSIS

- **Meiosis** involves two successive divisions (**meiosis I** & **meiosis II**), resulting in four cells (Figure 1.9).
- Each of these four cells has half the number of chromosomes of the mother cell and is genetically different.
- Meiosis produces cells (gametes) for sexual reproduction.
- It ensures constant chromosome number in sexually reproducing organisms (i.e. before fertilization chromosome number is halved).
- Each meiotic division proceeds through prophase, metaphase, anaphase, and telophase, but the behaviour of chromosomes is different compared with mitosis (Figure 1.10).
- Meiosis provides gametes for sexual reproduction, although meiosis does not always directly precede fertilization (see p. 21).
- Meiosis ensures constancy of chromosome number from generation to generation:
  - it generates gametes with half the species chromosome number
  - thus, at fertilization, the normal species chromosome number is re-established.

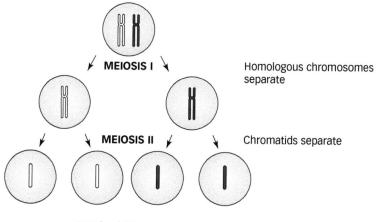

MEIOSIS I — Homologous chromosomes separate

MEIOSIS II — Chromatids separate

Four haploid gametes

Figure 1.9 The two divisions of meiosis (one pair of homologous chromosomes shown).

- Meiosis produces variability among progeny because it promotes new allele combinations in gametes by the:
  - ○ swapping of sections of homologous chromosomes during prophase 1
  - ○ **independent assortment** of pairs of homologous chromosomes at metaphase I.
- Random fertilization of these genetically different gametes further promotes offspring variability.
- Consider details of each stage described in Table 1.7 in conjunction with Figure 1.10.

Focus on prophase 1
- Prophase 1 is the longest and most critical phase of meiosis.
- Historically, it has been subdivided into five stages (Table 1.8). The key aspect is *not* the names of each stage, but the events that occur in each.
- The pivotal event is crossing over (Figure 1.11), which:
  - ○ involves the exchange of homologous sections of chromosomes between two non-sister chromatids
  - ○ produces hybrid patches of maternal and paternal chromosomes. It mixes the association of maternal and paternal alleles linearly along a chromosome, i.e. the alleles are **recombined**.

  ➲ *See section 4.1 (p. 60) for more details about recombination of alleles*
- Crossing over produces **variation**.
- It can occur several times between non-sister chromatids of one pair of homologous chromosomes, and will occur at different chromosomal sites during each meiosis.
- Chiasmata do not occur with equal probability along chromosomes.

Meiosis

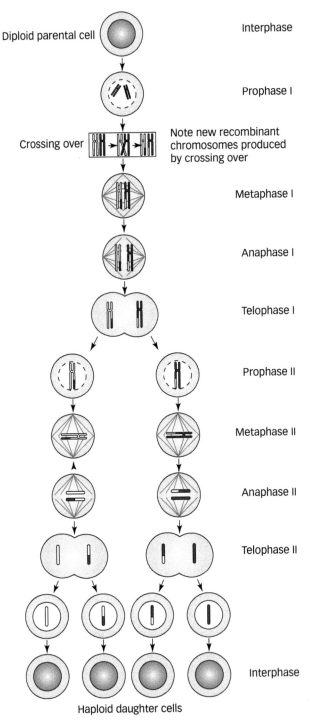

Figure 1.10 A summary of meiosis in an animal cell.

Stage	Key events
Prophase I	• Chromosomes condense • Homologous chromosomes **synapse** (i.e. pair) • **Crossing over** occurs between non-sister chromatids of homologous chromosomes • The nuclear membrane breaks down • Centrioles migrate to opposite poles and the spindle forms • Spindle fibres attach to kinetochores; one per pair of chromatids
Metaphase I	• Pairs of homologous chromosomes align on the spindle equator
Anaphase I	• Homologous chromosomes are pulled to opposite poles • Non-sister chromatids separate to opposite poles; sister chromatids migrate to same pole
Telophase I	• Nuclear envelopes form around the cluster of chromosomes at each pole • Cells may enter a brief interphase • No DNA replication before meiosis II
Prophase II	• Chromosomes condense • The nuclear membrane breaks down • A spindle forms and spindle fibres attach to kinetochores, one per chromatid
Metaphase II	• Pairs of chromatids align on the spindle equator, which is generally 90° to that of meiosis I
Anaphase II	• Sister chromatids are separated and pulled to opposite poles • Chromatids are, once again, termed chromosomes
Telophase II	• Nuclear envelopes form around the cluster of chromosomes at each pole • Chromosomes uncoil • Cytokinesis separates the nuclei into separate cells • Four cells result from single parental cell

Table 1.7 Details of each meiotic stage

Stage of prophase 1	Details of events
Leptonema	• Chromosomes appear as long thin threads; sister chromatids indistinguishable • Telomeres attach to nuclear membrane
Zygonema	• **Synapsis** (pairing of homologous chromosomes) occurs, facilitated by the **synaptonemal** complex—pairing is precise, often starting at the centromere and proceeding in a zipper-like fashion • Paired homologous chromosomes are termed **bivalents**
Pachynema	• Crossing over occurs to create a physical **chiasma(ta)**—chromosomes are now highly condensed, appearing short and thick
Diplonema	• Homologous chromosomes separate a little from each other • Chiasmata clearly visible
Diakinesis	• Chiasmata **terminalize**, i.e. move towards telomeres • Other prophase events occur: nuclear membrane disintegrates and spindle forms

Table 1.8 Events during prophase I

• For example, in humans more crossovers occur towards the telomeres of chromosomes.
• The homogametic sex in vertebrates has a higher chiasma count, e.g. crossovers is 1.5 times higher during human female meiosis (oogenesis) than male meiosis (spermatogenesis).

Meiosis

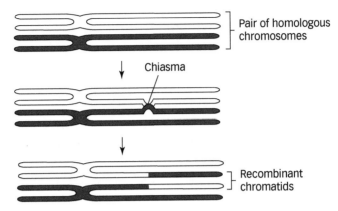

Figure 1.11 Crossing over between non-sister chromatids of a pair of homologous chromosomes.

Meiosis and life cycles

- Meiosis forms gametes for sexual reproduction, which only occurs in eukaryotes.
- Gametes are haploid; fertilization restores a species' diploid chromosome complement.
- Thus, sexually reproducing individuals alternate haploid and diploid phases in their life cycles.
- There are differences among phyla of different kingdoms on the length of time spent in each phase (Figure 1.12).
- Generally, **animals** have a short haploid phase. The products of meiosis develop into gametes with no intervening mitotic divisions.
- Many **plants** (e.g. mosses, ferns, flowering plants) show an **alternation** of two distinct generations in their life cycle. A diploid spore-producing **sporophyte** generation alternates with a haploid gamete producing **gametophyte** generation.
- Evolutionarily, the sporophyte generation has developed **dominance**, so flowering plants spend most of their existence as sporophytes, while the two generations are of equal length in mosses.
- Many **protists** and **fungi** have a haploid dominated life cycle.

⇒ *Check your understanding*

1.5 What mechanisms ensure cells begin mitosis with completely replicated chromosomes?

1.6 Describe the structure of a chromosome at metaphase of mitosis.

1.7 In what ways do mitosis and meiosis differ?

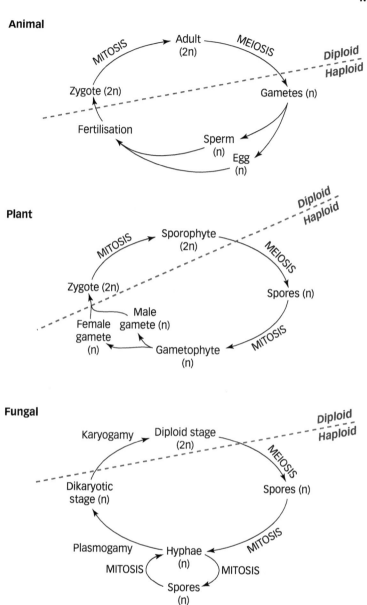

Figure 1.12 Generalized animal, plant, and fungal life cycles.

 online resource centre

You'll find guidance on answering the questions posed in this chapter—plus additional multiple-choice questions—in the Online Resource Centre accompanying this revision guide. Go to www. oxfordtextbooks.co.uk/orc/thrive/ or scan this image:

2 Principles of Mendelian Inheritance

Mendelian genetics is concerned with patterns of inheritance associated with one or a few genes.

Key concepts

- Monohybrid crosses investigate the genetic basis of traits determined by a single gene.
- The inheritance of two genes is followed in dihybrid crosses.
- Genes exist in alternative forms, or alleles, which are responsible for the expression of different forms of a trait.
- An individual possess two alleles for each gene; these may be similar or different.
- Alleles segregate into gametes during meiosis.
- Pairs of alleles, on separate chromosomes, segregate independently.
- The ratios of different types of progeny indicate the genotypes of parents.
- A chi-squared (χ^2)statistical test assesses whether observed progeny ratios match various expected progeny ratios.

2.1 THE MENDELIAN APPROACH

- Gregor Mendel (1822–1884) worked with the garden pea, *Pisum sativum*, which has a short generation time, is easy to manipulate, and produces large numbers of progeny—necessary characteristics for experimental genetic organisms.
- Mendel established a breeding procedure for analysing the genetic basis of a trait.
  - **True** or **pure-breeding** parents are used. This means each parent comes from a variety that only expresses one version of a character, e.g. only produces round or only produces wrinkled peas.
  - Parents of contrasting expression for a given trait are crossed, e.g. one parent comes from a strain producing round peas and the other from a wrinkled pea strain (Figure 2.1). This produces the **F1** (or **first filial**) generation.
  - F1 individuals are interbred, resulting in the **F2** (**second filial**) generation.
  - Numbers of individuals in the F1 and F2 generations expressing different versions of the trait are scored.
- A sufficient number of crosses are performed to produce a large number of progeny so that statistically significant patterns can be recognized.
- **Reciprocal crosses** are performed to eliminate gender-related expression, i.e. the traits expressed by the males and females are reversed. For example:
  - in 50% of the crosses the male parent produces round peas and the female produces wrinkled
  - in the other 50% of the crosses the female parent produces round peas and the male produces wrinkled.

2.2 MONOHYBRID CROSSES

- **Monohybrid crosses** involve parents that differ in the expression of a single trait.
- Figure 2.1 shows what Mendel observed in one of his monohybrid crosses involving plants that produced different shaped peas.

Figure 2.1 Investigating the inheritance of pea shape. (A) Producing the F1 generation. (B) Producing the F2 generation.

Monohybrid crosses

- We can use the results of Mendel's pea cross to establish key genetic concepts.
  - ◦ Traits are controlled by **genes**.
  - ◦ Genes exist in different versions or **alleles**. Thus, the pea shape gene has two alleles—one encoding production of round peas, the other wrinkled peas.
  - ◦ An individual possesses two alleles for a particular gene—one on each of a pair of homologous chromosomes (Table 2.1).
  - ◦ An individual can have:
    - (i) two 'round' alleles (pure-breeding 'round pea' parent)
    - (ii) two 'wrinkled' alleles (pure-breeding 'wrinkled pea' parent)
    - (iii) one 'round' and one 'wrinkled' allele (F1 individuals who have inherited one allele from each parent).
  - ◦ Alleles may be **dominant** or **recessive**. When both alleles for a trait are present, it is the dominant allele that is expressed.
  - ◦ Consider the F1 individual with two different alleles. It produces round peas; thus, the 'round' allele is dominant (see also Figure 2.2).

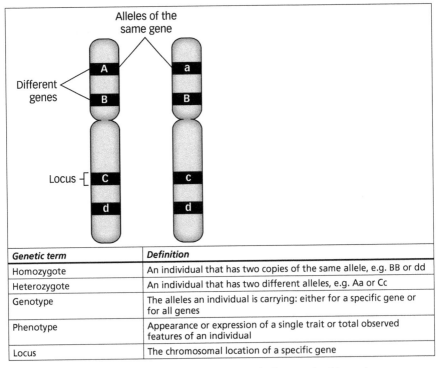

Genetic term	Definition
Homozygote	An individual that has two copies of the same allele, e.g. BB or dd
Heterozygote	An individual that has two different alleles, e.g. Aa or Cc
Genotype	The alleles an individual is carrying: either for a specific gene or for all genes
Phenotype	Appearance or expression of a single trait or total observed features of an individual
Locus	The chromosomal location of a specific gene

Table 2.1 Key genetic terms. The associated image depicts a pair of homologous chromosomes showing the relationship between genes, alleles, and chromosomes: four genes are shown

Looking for extra marks?

There is a biochemical explanation for the different pea shapes: the gene determining pea shape encodes a starch branching enzyme (SBE1). The SBE1 gene has two alleles: the dominant allele produces active SBE1, which produces high levels of amylopectin (the branched form of starch), while the recessive allele produces an inactive enzyme. In wrinkled peas (rr) there is no functional SBE1: amylopectin is at low levels, resulting in an irregular shaped seed which shrinks unevenly as it ripens.

 Check your understanding

2.1 What is the difference between a gene and an allele?

Representing genetic crosses

- Genetic cross diagrams illustrate inheritance details and aid understanding.
- Alleles are represented by symbols. Generally, the first letter of the dominant trait is chosen.
  - ○ The dominant allele is represented by the upper case letter and the recessive allele by the lower case letter.
  - ○ Considering the pea shape example, **R/r** represent the dominant round/ recessive wrinkled alleles.
- In a genetic cross diagram, the different possible gametes produced as a result of segregation of allele pairs during meiosis are shown. Likewise, all possible gamete combinations at fertilization are represented (Figure 2.2).

(A) Round pea producers	Wrinkled pea producers	*Pure breeding parents*
RR	rr	
R	r	**Possible gametes**
Rr All round pea producers		*F1*

(B) Round pea producers	Wrinkled pea producers	*F1 parents*
Rr	**Rr**	
R & r	**R & r**	**Possible gametes**
RR **Rr** Round pea producers	**Rr** rr Wrinkled pea producers	*F2*

Figure 2.2 Representing a genetic cross: crossing parents pure breeding for round or for wrinkled peas. (A) Production of F1 individuals. (B) Production of F2 individuals.

Monohybrid crosses

Phenotype	Genotype
Grey (dominant)	e+/e+ (homozygote) e+/e (heterozygote) (or sometimes represented as: +/+ and +/e)
Ebony (recessive)	e/e

Table 2.2 Representing *Drosophila melanogaster* body colour alleles

- An occasional alternative way of representing alleles is by:
  - a single letter with superscript + for the dominant allele
  - the same letter without the superscript for the recessive allele.
- This method is commonly used to represent genes of the fruit fly, *Drosophila melanogaster* (Table 2.2).

Punnett squares

- When an individual forms more than one gamete type it is best to represent the cross by a grid or **Punnett square**, rather than using lines (as in Figure 2.2) to show possible fertilizations.
- The use of a Punnett square reduces the chance of making a mistake when working out possible progeny allele combinations (Figure 2.3).
- In a Punnett square the gametes from one parent are displayed horizontally and from the other parent vertically.
- In Figure 2.3 four equally likely gamete combinations arise illustrating that there is a:
  - 25% chance of **RR** plants
  - 50% chance of **Rr** plants
  - 25% chance of **rr** plants.
- Or, expressed phenotypically, 75% of plants producing round peas and 25% producing wrinkled peas are expected from a cross between two **heterozygous** round pea producers.
- If parental genotypes are known a Punnett square enables prediction of phenotypic expression in future generations.

Gametes	R	r
R	RR round pea producers	Rr round pea producers
r	Rr round pea producers	rr wrinkled pea producers

Figure 2.3 Using a Punnett square to predict results of crossing pea plants heterozygous for producing round seeds (**Rr**).

The Law of Segregation

- The **Law of Segregation** (*Mendel's First Law*) states that the two members of an allele pair separate from each other into different gametes.
- Events during meiosis explain segregation of alleles into gametes.
- Consider one gene with two alleles (**A/a**). The gene will be found at a particular **locus** on each of a pair of homologous chromosomes (Figure 2.4).
- Each allele segregates into separate cells during meiosis because:
  - during meiosis I pairs of homologous chromosomes separate
  - during meiosis II pairs of chromatids separate.
- The segregation of alleles during gamete production and subsequent random association in new combinations with alleles of other genes at fertilization ensures variation among offspring.

The test cross

- The **test cross** is a useful way of identifying the genotype of an individual who shows a dominant phenotype.
- It involves crossing an individual showing a dominant phenotype with another individual who is **homozygous** for the recessive allele (Figure 2.5).
  - If the 'test' individual is homozygous dominant all test cross progeny will also show the dominant phenotype (they will be heterozygous).

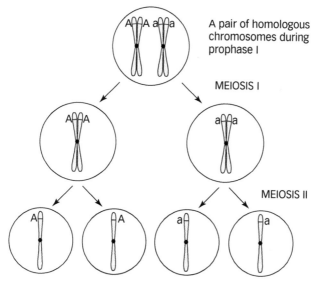

Four haploid gametes; half received allele 'A' and the other half received allele 'a'

Figure 2.4 Meiosis segregates pairs of alleles into separate gametes.

Dihybrid crosses

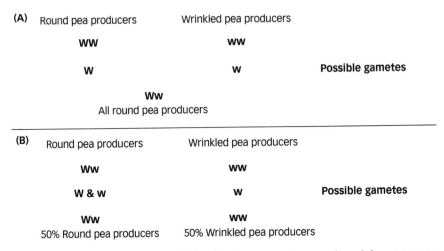

(A) Round pea producers | Wrinkled pea producers

WW | ww

W | w | **Possible gametes**

Ww
All round pea producers

(B) Round pea producers | Wrinkled pea producers

Ww | ww

W & w | w | **Possible gametes**

Ww | ww
50% Round pea producers | 50% Wrinkled pea producers

Figure 2.5 Results of a test cross. (A) The dominant round pea producer is homozygous. (B) The dominant round pea producer is heterozygous.

- ○ If the 'test' individual is heterozygous an equal number of dominant and recessive progeny (a 1:1 ratio) results.
- A test cross is sometimes referred to as a **backcross**, as the genotype testing of individuals showing a dominant phenotype often occurs in the context of genetic crosses, when dominant expressing individuals are 'back' crossed to an original homozygous recessive parent.

Check your understanding

2.2 In guinea pigs short hair is dominant to long hair. Using symbols S for the short hair allele and s for the long hair allele write the genotypes of (i) a heterozygous guinea pig; (ii) a long hair guinea pig; and (iii) guinea pigs produced by crossing pure-breeding short and pure-breeding long hair guinea pigs. (iv) What is the phenotype of a guinea pig with genotype SS?

2.3 DIHYBRID CROSSES

- **Dihybrid crosses** consider the inheritance patterns produced by the segregation of alleles of two genes.
- In a dihybrid cross the two genes may:
  - ○ each determine a separate trait
  - ○ together influence a single trait.

Parents:	Pure breeding normal wings / red eyes	Pure breeding vestigial wings / sepia eyes

F1 generation: All normal wings / red eyes

F1 generation: 9 normal wings / red eyes
3 normal wings / sepia eyes
3 vestigial wings / red eyes
1 vestigial wings / sepia eyes

Figure 2.6 Mendelian analysis of the genetic basis inheritance of two traits (normal/vestigial winged and red/sepia eyes) of the fruit fly, *Drosophila melanogaster*

- The standard Mendelian approach to genetic analysis (described in section 2.1) is used to identify the involvement of two genes, i.e.:
  - pure-breeding parents of contrasting expression for the trait(s) being investigated are crossed to produce an F1 generation
  - F1 individuals are interbred to produce the F2 generation
  - proportions of individuals in the F1 and F2 generations expressing different combinations of the trait(s) are scored. These proportions indicate the genetic basis of the trait(s).
- Figure 2.6 represents a dihybrid cross.
- Considering the results of the breeding experiment shown in Figure 2.6:
  - the F1 phenotypes indicate that normal wings and red eyes are the dominant traits
  - the F2 phenotypic ratio of 9:3:3:1 indicates that two genes are involved, one controlling each character.
- Figure 2.7 repeats the breeding experiment shown in Figure 2.6, but shows genotypes.
- Using Figure 2.7 as a representative example of a dihybrid cross, note how:
  - each gamete has one allele for each trait
  - each F1 individual produces four kinds of gametes in equal proportions—these result from the independent segregation of each allele pair
  - gametes combine at random—as any 1 of the 4 gametes of one parent can combine with any 1 of the 4 gametes of the other parent this produces 16 possible fertilization outcomes
  - F2 generation shows phenotypes in predictable proportions:
    - 9/16 dominant for both traits
    - 3/16 dominant for one trait and recessive for other
    - 3/16 recessive for first trait and dominant for second
    - 1/16 recessive for both traits.

  ➔ *See section 3.6 for more dihybrid crosses.*

Dihybrid crosses

Parental phenotypes:	Normal wings / red eyes	Vestigial wings / sepia eyes
Genotypes:	NNRR	nnrr
Gametes:	NR	nr

F1 genotype: NnRr
F1 phenotype: Normal wings / red eyes

Gametes	NR	nR	Nr	nr
NR	**NNRR** normal / red	**NnRR** normal / red	**NNRr** normal / red	**NnRr** normal /red
nR	**NnRR** normal / red	**nnRr** vestigial / red	**NnRr** normal / red	**nnRr** vestigial / red
Nr	**NnRr** normal / red	**NnRr** normal / red	**NNrr** normal / sepia	**Nnrr** normal / sepia
nr	**NnRr** normal / red	**nnRr** vestigial / red	**Nnrr** normal / sepia	**Nnrr** vestigial / sepia

F2 genotype:	N–R–	nnR–	N–rr	nnrr
F2 phenotype:	Normal red 9	Vestigial red 3	Normal sepia 3	Vestigial sepia 1

Figure 2.7 F1 and F2 genotypes and phenotypes of the dihybrid cross between pure-breeding normal-winged/red-eyed and vestigial-winged/sepia-eyed parents. **N/n** represent alleles for normal/vestigial wings and **R/r** represent alleles for red/sepia eyes.

The Law of Independent Assortment

- **The Law of Independent Assortment** (*Mendel's Second Law*) states that different pairs of alleles assort independently of each other at gamete formation.
- Events during meiosis explain the independent assortment of pairs of alleles.
- Different pairs of homologous chromosomes orientate themselves randomly on the equator during metaphase I.
- Consider two genes each with two alleles (**A/a** and **B/b**) found at specific loci on different pairs of homologous chromosomes. Because of random orientation of pairs of homologous chromosomes on the equator during metaphase I, one member of each allele pair is equally likely to appear in a gamete with either of the two alleles of the other pair, i.e. allele **A** and allele **a** are equally likely to be partnered in a gamete with allele **B** or **b** (Figure 2.8).
- Thus, independent assortment combined with random fertilization of gametes produces variability among offspring.

Dihybrid test cross

- If an individual expressing the dominant phenotype for two traits is thought to be heterozygous for both, crossing this individual with another that is homozygous recessive for both traits will produce four phenotypes in a 1:1:1:1 ratio (Figure 2.9).

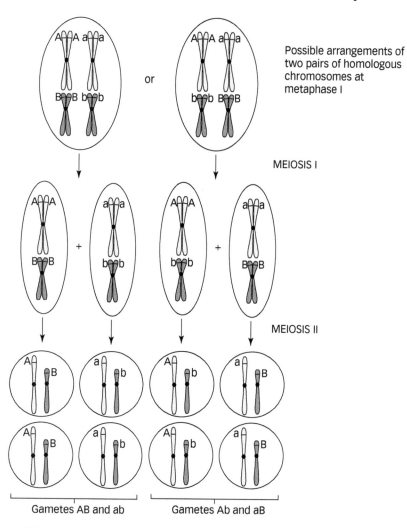

Possible arrangements of two pairs of homologous chromosomes at metaphase I

MEIOSIS I

MEIOSIS II

Gametes AB and ab   Gametes Ab and aB

Figure 2.8 Independent assortment during meiosis in a heterozygote of genotype AaBb. This produces gametes of four possible genotypes. Note that each individual meiosis produces only two types of gametes with respect to allele combinations at two loci on separate chromosomes.

Mendelian cross summary

- There is a relationship between the number of independently segregating loci and the number of different gametes, genotypes, and phenotypes:
  - n = number of heterozygous loci
  - 2^n = gamete types
  - 3^n = F2 genotypes
  - 2^n = phenotypes (complete dominance at all loci).

Dihybrid crosses

Test fly

Normal wings / red eyes   Vestigial wings / sepia eyes
NnRr   nnrr

Gametes   NR Nr nR nr   nr

Gametes	NR	Nr	nR	nr
nr	NnRr Normal / red	Nnrr Normal / sepia	nnRr Vestigial / red	Nnrr Vestigial / sepia

Progeny show phenotypes in predictable proportions:
¼ normal/red - *dominant for both traits*
¼ normal/sepia - *dominant for first trait/recessive for second*
¼ vestigial/red - *recessive for first trait/dominant for second*
¼ vestigial/sepia - *recessive for both traits*

Figure 2.9 A dihybrid test cross: a fruit fly showing normal wings and red eyes (dominant phenotype) is crossed with another homozygous for vestigial wings and sepia eyes.

- Thus, only a few parental differences can lead to great progeny variation:

  parents differ at one locus—F2 genotypes = 3 genotypes (Table 2.2)

  parents differ at two loci—F2 genotypes = (3×3) 9 genotypes (Figure 2.8)

  parents differ at three loci—F2 genotypes = (3×3×3) 27 genotypes (Figure 2.11).

- It is important to remember that many characteristics of living organisms are also influenced by non-genetic factors, which increases variable expression of a trait.

  ⮕ *See Chapter 5 for details of traits influenced by both genetic and non-genetic factors.*

Looking for extra marks?

A trihybrid cross involves simultaneously considering the inheritance of three genes. If each gene has two possible alleles, gametes produced by true-breeding parents (AABBCC) and (aabbcc) will be ABC and abc. The resulting F1 heterozygote (AaBbCc) produces eight different gametes. If two F1 heterozygotes are interbred, 64 different fertilizations and 27 different genotypes arise.

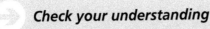 Check your understanding

2.3 Papermouths are a genus of freshwater fish. They may be seven- or five-spined, and dark or pale pigmented. A fish farmer crossed 2 seven-spined, dark pigmented fish and obtained 56 seven-spined dark pigmented fish, 16 seven-spined pale pigmented fish, 21 five-spined dark pigmented and 6 five-spined pale pigmented fish. What can you deduce about the inheritance of these two traits?

2.4 PROBABILITY AND MENDELIAN GENETICS

- **Probability** may be defined as the proportion of times that a specific outcome occurs in a series of events.
- For example, the probability that a couple's child will be a son is 0.5.
- Because probabilities are proportions they lie numerically between 0 and 1.
- Probability plays a central role in Mendelian genetics because it:
  - aids interpretation of the outcome of genetic crosses
  - enables predictions to be made of future progeny genotypes and therefore phenotypes.
- In applying probability rules to Mendelian genetics it is important to appreciate the chance nature of key events during meiosis and fertilization.
- During meiosis the outcome of two chance events—(i) crossing over at prophase I and (ii) independent assortment of pairs of homologous chromosomes—produces gametes with different combinations of alleles.
- Furthermore, it is a random event as to which gametes fuse during fertilization.
- The two useful rules of probability in Mendelian genetics are the **multiplication rule** and the **addition rule**.

The multiplication rule

- The probability of the co-occurrence of two (or more) independent events is obtained by multiplying the probabilities of each independent event.
- Consider the possible genotypes and phenotypes of a child conceived by parents who are each heterozygous (**Aa**) for albinism (Table 2.3).

The addition rule

- The probability of one event *or* another event occurring is calculated by adding the probabilities of each event, assuming the two different events occur independently of each other.

Gametes	A	a
A	AA Normal pigmentation	Aa Normal pigmentation
a	Aa Normal pigmentation	aa Albino

- Parents may want to know the probability that three successive children will be albino.
- This probability is calculated as follows:
  - the probability of a child expressing albinism = ¼
  - the probability that three successive children express albinism is

$$\tfrac{1}{4} \times \tfrac{1}{4} \times \tfrac{1}{4} = \tfrac{1}{64}$$

Table 2.3 Predicting the probability of expressing albinism. **A** represents the allele for normal pigmentation; **a** is the allele for lack of pigmentation

- Consider again the inheritance of albinism (Table 2.3). A couple know that they are both carriers for albinism. They wish to have two children and want to know the probability that one is albino.
- For this couple, having an albino child could occur in one of two ways:
  - first child albino/second child normal pigmentation
  - first child normal pigmentation/second child albino.
- The probability of each possible scenario is:

$$\frac{1}{4} \times \frac{3}{4} = \frac{3}{16}$$
$$\frac{3}{4} \times \frac{1}{4} = \frac{3}{16}$$

- Final probability = probability of pattern 1 + probability of pattern 2

$$= \frac{3}{16} + \frac{3}{16}$$
$$= \frac{6}{16} \text{ or } \frac{3}{8}$$

i.e. if the couple have 2 children, there is a 3/8 probability that 1 will be albino.

Check your understanding

2.4 Phenylketonuria (PKU) is an autosomal recessive human disease that is characterized by a build up of the amino acid phenylalanine in the blood. If undetected at birth it can lead to mental retardation. Jess and Alex are both heterozygous for the PKU gene. Jess is pregnant with twins. What is the probability both twins will suffer from PKU if (i) they are dizygous (non-identical) twins; (ii) they are monozygous (identical) twins?

2.5 THE CHI-SQUARED TEST AND MENDELIAN GENETICS

- The chi-squared (χ^2) statistical test is used to determine whether or not chance events can reasonably explain a deviation of observed data from the hypothesized result.
- It is used in Mendelian genetics to compare observed progeny numbers with expected ratios, because the ratios of different progeny phenotypes can be informative of underlying genetics.
- We need to know whether observed deviations from expected ratios:
  - are the result of random fluctuations arising from the chance nature of fertilization, and the segregation and independent assortment of alleles during meiosis
  - indicate an unexpected genetic explanation.

	Yellow fruit virus resistant	Yellow fruit virus susceptible	White fruit virus resistant	White fruit virus susceptible
Observed number of plants	63	49	59	45

Table 2.4 Results of a test cross between melons with yellow skins/resistant to yellows virus and melons with white fruit/susceptible to yellows virus

- Consider the example of a melon breeder who doubts his dominantly-expressing plant is homozygous for yellow skin and resistance to yellows virus. He performs a test cross of his dominant plant with a plant homozygous recessive for these traits. Table 2.4 shows the plants produced from the resulting seeds
- The results presented in Table 2.4 suggest a 1:1:1:1 phenotypic ratio, which indicates the melon grower's dominantly-expressing plant is heterozygous. Not what he wanted! A χ^2 test confirmed the 1:1:1:1 phenotypic ratio (Box 2.1).
  ➔ *See section 2.3, p. 30 for details of dihybrid test crossing.*

BOX 2.1 Use of the chi-squared statistical test to verify test cross phenotype ratio

- The chi-squared (χ^2) statistical test compares observed (O) and expected (E) data:

$$\chi^2 = \sum \frac{(O-E)^2}{E}$$

- For each frequency component (here phenotype class):
  - (i) the expected value is subtracted from the observed value
  - (ii) the resulting value is squared and then divided by the expected value
  - (iii) the values for each frequency component are summed to produce a **calculated χ^2 value**
  - (iv) the calculated value is compared with the critical χ^2 value
  - (v) the critical χ^2 value depends on the degrees of freedom (number of data categories minus 1) and significance level (probability that null hypothesis is correct, generally 0.05 for genetic data)
  - (vi) the null hypothesis is accepted if the calculated value is less than the critical χ^2 value.
- Null hypothesis: there is no difference between observed and an expected ratio of 1:1:1:1.
- At a significance level of 0.05 and 3 degrees of freedom, the critical χ^2 value = 7.82. The calculated χ^2 value (3.92) is less than the critical value, thus the null hypothesis is accepted. The results therefore resemble a 1:1:1:1 ratio.
- Deviations were due to chance non-genetic factors. The yellow-skinned, virus-resistant plant is a double heterozygote.

The chi-squared test and Mendelian genetics

	Observed numbers (O)	Expected numbers (E)	O – E	(O – E)²	$\frac{(O-E)^2}{E}$
Yellow resistant	63	54	9	81	1.5
Yellow susceptible	49	54	–5	25	0.46
White resistant	59	54	5	25	0.46
White susceptible	45	54	–9	81	1.5
	216	216			3.92

Table 2.5 Producing a calculated χ^2 value from the test cross data of Table 2.4

Check your understanding

2.5 What chromosomal events during meiosis ensure that (i) pairs of alleles segregate into separate gametes; (ii) pairs of alleles independently assort?

2.6 You have pure breeding black-eyed mice and pure-breeding red-eyed mice. What crosses would you perform to investigate the genetic basis of mouse eye colour. What would you be looking for among the progeny?

2.7 What is a test cross? How is a χ^2 statistical test used to analyse the results of a dihybrid test cross?

2.8 How many different genotype classes would you expect in a trihybrid test cross (involving genes on separate chromosomes)? What would these genotype classes be and how would they relate to phenotypes?

online resource centre You'll find guidance on answering the questions posed in this chapter—plus additional multiple-choice questions—in the Online Resource Centre accompanying this revision guide. Go to www. oxfordtextbooks.co.uk/orc/thrive/ or scan this image:

3 Further Mendelian Principles

Modified Mendelian inheritance patterns indicate various genetic interactions and influencing factors.

Key concepts

- Genes and their alleles are always inherited according to Mendelian principles, but the phenotypic expression of a trait does not always follow simple Mendelian patterns.
- Different phenotypic ratios occur among F2 progeny in Mendelian breeding experiments when:
  ○ alleles are incompletely or co-dominant
  ○ epistatic interactions occur between genes
  ○ an allele is located on a sex chromosome or organelle chromosome.
- There may be more than two alleles possible at a given locus.
- A particular combination of alleles may be lethal.
- An understanding of varied Mendelian patterns of expression is gained from relating phenotypes to underlying molecular processes and interactions.
- The genetic basis of many human traits is deduced from studying inheritance patterns through succeeding generations in pedigrees.

3.1 INCOMPLETE AND CO-DOMINANCE

Incomplete dominance

Incomplete dominance occurs when heterozygous individuals show a phenotype intermediate between the phenotype of each homozygote (Table 3.1).

- When two alleles show incomplete dominance, the intermediate phenotype of the heterozygote can often be related to functional enzyme levels.
- For example, consider flower colour in snapdragons, *Antirrhinum major*. Crossing pure-breeding red parents (RR) with pure-breeding white parents (rr) produces F1 plants with pink (Rr) flowers (Figure 3.1).
- The R allele encodes a functional enzyme in the red pigment production pathway. No functional enzyme is produced in plants with r allele.
- Functional enzyme levels are reduced in the F1 heterozygote. Thus, only a proportion of precursor white molecules are converted to the red pigment. A mixing of the red and white molecules produces the intermediate pink colour.
- If F1 plants with pink petals are interbred, red-, pink-, and white-petalled plants result in a ratio of 1:2:1 (Table 3.2). This is an important diagnostic Mendelian ratio.

Genotype	Complete dominance	Incomplete dominance
AA	Phenotype 1	Phenotype 1
Aa	Phenotype 1	Phenotype 2
aa	Phenotype 2	Phenotype 3

Table 3.1 A comparison of phenotypic expression when the two alleles (A/a) of a gene are completely and incompletely dominant

Phenotype:	Red petals	White petals
Genotype:	RR	rr
Gametes:	R	r
F1 genotype:	Rr	
F1 phenotype:	Pink petals	

Figure 3.1 The inheritance of snapdragon petal colour.

Gametes	R	r
R	RR Red petals	Rr* Pink petals
r	Rr Pink petals	rr White petals

Table 3.2 Interbreeding heterozygous pink-petalled snapdragons (Rr × Rr)

*When two alleles show incomplete dominance they are sometimes represented by two upper case letters, e.g. R/W for red/white petals. So, snapdragons producing pink flowers are also represented by genotype RW.

Squirrel phenotype	Squirrel genotype	% of MC1R responsive to hormone
Grey	mm	100
Intermediate: brown/ black	Mm	50
Black	MM	0

Table 3.3 The genetic basis of pigmentation in the grey squirrel

- Incomplete dominance of alleles of genes coding for molecules other than enzymes can produce heterozygotes of intermediate phenotype.
- For example, the UK grey squirrel (*Scurius carolinensis*) has a melanic (black) variant produced by a mutation in a cell membrane receptor—the melanocortin receptor (MC1R). The mutation abolishes sensitivity to a regulatory hormone (Table 3.3).

Co-dominance

- **Co-dominance** is the simultaneous expression of both homozygous traits by the heterozygote.
- This is well illustrated by the human ABO blood groups, determined by the nature of a cell surface oligosaccharide antigen (**blood group antigen**) on human erythrocytes.
- Three alleles exist for the ABO gene, which encodes a transferase enzyme catalysing the addition of a sugar to a precursor five-sugar H antigen (Table 3.4).
- Depending on the antigens present in the erythrocyte membranes, four blood groups result (Table 3.5).
- The different ABO blood groups are further characterized by their phenotypic response to blood transfusions (Table 3.6).

ABO allele (enzyme produced)	Enzyme function (what the enzyme adds to the precursor H antigen)	Blood group antigen
A (A-transferase)	N-acetylgalactosamine	A
B (B-transferase)	Galactose	B
O (no enzyme)	No sugar	None

Table 3.4 ABO blood group alleles and their functional products

Phenotype blood group (antigen present)	Genotype
A (A antigen)	AA or AO
B (B antigen)	BB or BO
AB (A & B antigens)*	AB
O (neither)	OO

Table 3.5 ABO blood groups
*Co-dominance: the heterozygote (AB) expresses the traits of both homozygotes (AA and BB).

Recipient's blood group	Donor blood group
A	A or O
B	B or O
AB	AB, A, B or O
O	O

Table 3.6 Blood transfusions: compatible donor and recipient ABO blood groups

Looking for extra marks?

Production of the H antigen (the precursor of the A and B blood group antigens) is catalysed by the enzyme fucosyl transferase, which is encoded by the *FUT1* gene. Mutation in this gene produces a pseudo-O blood group phenotype, the **Bombay phenotype** (designated O_h).

Check your understanding

3.1 A boy is blood group O. What blood group could his mother not be?

3.2 MULTIPLE ALLELES

- A gene can have more than two (i.e. multiple) alleles.
- A single diploid individual can only possess two of many possible alleles for a gene.
- Different **multiple alleles** are maintained at varying frequencies within a population.

  ➡ *See section 13.1 (p. 252) for calculation of allele frequencies in different populations.*
- New alleles are created by changes in DNA sequence, i.e. mutation.
- Multiple alleles often show a hierarchy of dominance relationships referred to as a **dominance series**.
- An example of a dominance series are the four common alleles of the rabbit coat colour gene: agouti/chinchilla/himalayan/albino (Table 3.7), with dominance in that order.
- Many human disease genes possess multiple recessive alleles, e.g. there are nearly 2000 'alleles' of the human cystic fibrosis gene.
- Alleles of human disease genes are generally termed 'mutations'. Each produces a different version of the encoded protein.

Phenotype	Genotype
Agouti (the brown-grey of wild rabbits)	CC Cc^{ch} Cc^h Cc
Chinchilla (light grey)	c^{ch}c^{ch} c^{ch}c^h c^{ch}c
Himalayan (white with black extremities)	c^{h}c^h c^{h}c
White	cc

Table 3.7 The genetic control of rabbit coat colour

Allele symbols: C = agouti; C^{ch} = chinchilla; C^h = Himalayan; c = albino.

- An expressing homozygous recessive individual often has two different recessive alleles and can also be described as a **compound heterozygote**.
- As different mutations affect the encoded protein in different ways, the severity of a disease can vary in different compound heterozygotes.

> ### Looking for extra marks?
>
> Many flowering plants are hermaphrodite and often use genetic self-incompatibility to prevent self-fertilization. Typically, there is a self-sterility gene (S) with multiple alleles. Pollen grains (haploid) have a single S allele which will only develop a pollen tube if this allele is different to the two S alleles of the female stigma and style (diploid) tissues.

3.3 LETHAL ALLELES

- **Lethal alleles** eliminate a function essential to survival.
  - Recessive lethal alleles kill individuals that are homozygous for the allele.
  - Dominant lethal alleles kill both heterozygotes and homozygotes.
- Lethal alleles are detected by the deviations they cause to expected Mendelian ratios.
- When a lethal allele is expressed the embryo often dies *in utero,* resulting in the absence of an expected phenotypic class of progeny.
- When crossing two heterozygotes, with complete dominance of one allele over the other, the expected dominant to recessive phenotypic ratio among the progeny is 3:1. When the recessive allele is lethal in the homozygote, the progeny phenotypic ratio is 2:1.
- An often quoted lethal allele example is the yellow coat colour allele in mice (Table 3.8).
- When heterozygous yellow mice, $A^w A^y$, are crossed 1/3 of progeny are agouti and 2/3 are yellow (Table 3.9).
- The differing phenotypes resulting from genotypes $A^w A^y$ and $A^y A^y$ have a molecular explanation. Allele A^y has a large deletion which includes two genes: one determining coat colour and the other (*Merc*) influencing development. In the heterozygote, A^y affects coat colour, but there is a functional *Merc* product associated with the A^w allele. In the homozygote ($A^y A^y$) there is no *Merc* product, so development fails.

Genotype	Phenotype
$A^w A^w$	Agouti (**wild type**)
$A^w A^y$	Yellow (A^y acts as a dominant allele in the heterozygote)
$A^y A^y$	Lethal (A^y acts as a recessive allele in the homozygote)

Table 3.8 Genetic determination of mouse coat colour. Normal (agouti) coat colour is determined by allele A^w; yellow coat colour allele is A^y

Gametes	A^w	A^y
A^w	$A^w A^w$ (agouti)	$A^w A^y$ (yellow)
A^y	$A^w A^y$ (yellow)	$A^y A^y$ (lethal)

Table 3.9 The genetic consequences of mating two yellow mice ($A^w A^y$)

- Another example of a recessive lethal allele is the tail determining gene (T/t) of Manx cat. *TT* cats have tails, but *Tt* cats lack tails (described as 'Manx'). The *tt* genotype is lethal, so embryos die *in utero*.
- It is difficult to identify dominant lethal alleles as they exert their effect in both heterozygotes and homozygotes.
- Dominant lethal alleles can be identified when their action is delayed, e.g. the dominant allele causing expression of the human neurodegenerative Huntington disease.
- Symptoms of Huntington disease generally begin around the age of 30 years or later.

Check your understanding

3.2 How does phenotypic expression of the heterozygote differ in complete dominance, incomplete dominance, and co-dominance?

3.3 How are lethal alleles maintained in a population's **gene pool**?

3.4 SEX-LINKED TRAITS

- **Sex-linked traits** are encoded by genes on a sex chromosome.
  - ⊛ *See section 1.3 for details of sex chromosomes.*
- Males and females show different patterns of inheritance and expression of sex-linked traits, in contrast with the expression of genes on autosomes for which both sexes have equal probabilities of inheriting and expressing traits.

Expression of X-linked genes

- Males (XY) are **hemizygous** for X-linked genes, i.e. they only have one copy.
- Females (XX) can be homozygous or heterozygous for alleles of X-linked genes of which there are two copies.
- Recessive **X-linked traits** are expressed more often in males than females. This is because males are hemizygous. Thus, if they possess one recessive allele, then they express the recessive condition (Figure 3.2).
- When representing the genotype of an X-linked gene it is best to show the allele symbol alongside an X and to include the Y chromosome (Figure 3.2). This reduces the chance of errors when working out the results of genetic crosses.
- As with all genetic crosses a Punnett square can be used to work out possible outcomes of certain matings (Table 3.10).

Female	Genotypes:	X^HX^H	X^HX^h	X^hX^h
	Phenotypes:	normal clotting	normal clotting	haemophilia

Male	Genotypes:	X^HY	X^hY
	Phenotypes:	normal clotting	haemophilia

Figure 3.2 The possible male and female genotypes, and associated phenotypes for the human X-linked condition of haemophilia A. Allele H produces normal blood clotting factor VIII; allele h denotes lack of functional clotting factor VIII

Sex-linked inheritance patterns

- Reciprocal crosses give different results. This is a useful clue that a gene is located on a sex chromosome.
- A Mendelian-type reciprocal cross of red- and white-eyed *Drosophila melaonogaster* flies is illustrated in Figure 3.3.
- Considering the cross represented in Figure 3.3:
  - the F1 results clearly show that red-eyed is dominant
  - the differing F1 and F2 results, which depend upon which parent was white- or red-eyed, indicate a sex-linked gene

Gametes	X^H	X^h
X^H	$X^H X^H$ blood clots normally	$X^H X^h$ blood clots normally
Y	X^HY blood clots normally	X^hY haemophiliac*

*A probability of 0.25 that a child will be a haemophiliac and of 0.5 that a son is a haemophiliac.

Table 3.10 Using a Punnett square to assess the probability of children expressing haemophila when the mother carries the haemophilia A allele (X^HX^h) and the father's blood clots normally (X^HY)*

	Cross 1		Cross 2	
Parents	Red-eyed female	White-eyed male	Red-eyed male	White-eyed female
	X^RX^R	X^rY	X^RY	X^rX^r
F1 result	X^RX^r	X^RY	X^RX^r	X^rY
	Red-eyed females and males		Red-eyed females White-eyed males	
F2 result	Females: 50% X^RX^R (red-eyed)		Females: 50% X^RX^r (red-eyed)	
	50% X^RX^r (red-eyed)		50% X^rX^r (white-eyed)	
	Males: 50% X^RY (red-eyed)		Males: 50% X^RY (red-eyed)	
	50% X^rY (white-eyed)		50% X^rY (white-eyed)	

Figure 3.3 Investigating the genetic basis of *Drosophila melaonogaster* eye colour. Pure-breeding parents are crossed to give the F1 generation, which are interbred to produce F2 flies.

Sex-linked traits

Gametes	X^R	X^r
X^R	$X^R X^R$ red-eyed females	$X^R X^r$ red-eyed females
Y	X^R Y red-eyed males	X^r Y white-eyed males

Table 3.11 Using a Punnett square to identify possible F2 progeny from an F1 cross (cross 1) in Figure 3.3

- ○ males that inherit a single recessive allele express the recessive trait, as they are hemizygous; females need to be homozygous recessive to develop white eyes.
- As with all genetic crosses a Punnett square can be used to work out results (Table 3.11).

Z-linked inheritance in birds

- As females are the heterogametic sex (ZW) and males the homogametic one (ZZ), the pattern of sex-linked inheritance is *reversed*.
- Females express sex-linked traits more frequently. Males are the carriers.
- For example, green/yellow plumage in canaries is sex (Z)-linked. As green plumage is dominant, only female yellow canaries can be produced from crossing green canaries (Table 3.12).

Gametes	Z^G	Z^g
Z^G	$Z^G Z^G$ green male	$Z^G Z^g$ green male
W	Z^GW green female	Z^gW yellow female

Table 3.12 Using a Punnett square to identify possible progeny from a cross between green female and heterozygous green male canaries

Y-linkage

- **Y-linkage** refers to a trait determined by a gene on the Y chromosome; it is also known as holandric inheritance.
- Hairy ears seem to follow a male-to-male transmission that suggests Y-linkage.

Sex-influenced and sex-limited traits

- Sex-influenced and sex-limited traits are controlled by genes on autosomes.
- They are expressed differently in males and females due to different sex hormonal environments. The distinction is subtle:
  - (i) **sex-limited traits** are expressed in only one sex, e.g. increased susceptibility to breast cancer caused by mutant alleles of *BRCA1* occurs only in females, and cryptorchidism (undescended testicles) is only expressed in males
  - (ii) **sex-influenced traits** occur in both sexes, but more commonly in one, e.g. pattern baldness (when hair loss spreads out from the crown of the head) (Table 3.13).

Genotype	Male phenotype	Female phenotype
BB	Pattern baldness	Pattern baldness
Bb	Pattern baldness	Normal
bb	Normal	Normal

Table 3.13 The expression of pattern baldness in males and females

 Check your understanding

3.4 How do sex-limited and sex-influenced traits differ from sex-linked traits?

3.5 Feather colour in chickens is a sex-linked trait. Explain the following cross: Delaware (white with black barring on wings, neck, and hackles) females crossed with Rhode Island red males resulted in Delaware white males and Rhode Island red females.

3.5 EXTRACHROMOSOMAL INHERITANCE

- Mitochondria and chloroplasts contain their own genetic systems.
- Mutations in mitochondrial and chloroplast genes produce non-Mendelian inheritance patterns.
- The results of reciprocal crosses are different (Table 3.14).
- This is because males and females do not contribute mitochondria and chloroplasts equally to the next generation.
- Most/all of the next generation's cytoplasm comes from the female gamete.
- Thus, mitochondrial and chloroplast genomes show **maternal inheritance**, i.e. progeny show the phenotype of the female parent.

➔ *See section 6.7 (p. 120) for details of genetic systems of organelles.*

Female phenotype	Male phenotype	Progeny phenotype
Green	Green, variegated or white	Green
Variegated*	Green, variegated or white	Variegated
White	Green, variegated or white	White

*Variegated leaves have irregular patches of green and white colour.

Table 3.14 Inheritance of leaf colour in the four o'clock plant, *Mirabilis japonica*

Looking for extra marks?

Maternal inheritance of organelle DNA is universal. Well-known fungal examples are the respiratory *poky* and *petite* mitochondrial mutations of *Neurospora crassa* and *Sacchromyces cerevisiae* respectively. Mutations in the 16S rRNA of the chloroplast genome of the unicellular green alga *Chlamydomonas reinhardtii* confers streptomycin resistance.

Maternal influenced traits

- Nuclear genes encode **maternal influenced traits.**
- Thus, alleles determining these traits are inherited from both parents.
- However, an individual's expression of a maternal influenced trait is determined by the mother's genotype.
- The maternal influence arises because the products of the gene:
  - are needed early in development
  - are the products of maternal genes that were deposited in the egg cytoplasm prior to fertilization.
- Shell coiling in the freshwater snail *Lymnaea peregra* is a maternal influenced trait.
- The direction of coiling is determined by a nuclear gene.
- The dextral (s^+), right coiling allele is dominant to the sinistral (s), left coiling allele.
- Figure 3.4 illustrates how an individual snail may have a dominant genotype with respect to coiling, yet expresses the recessive phenotype, as the maternal genotype was recessive.

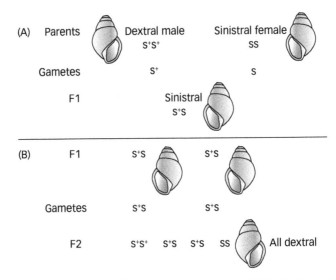

Figure 3.4 Inheritance of shell coiling in *Lymnaea peregra*. (A) Production of the F1 generation: F1 snails are heterozygous with a dominant dextral allele; they express the maternal phenotype. (B) Production of the F2 generation: all snails are right coiling (maternal genotype), even though 25% of F2 snails are homozygous sinistral.

Check your understanding

3.6 How could you determine whether a trait was due to maternal inheritance or maternal influenced inheritance?

3.6 EPISTASIS

- **Epistasis** is a situation in which the products of two or more genes interact to influence expression of a single trait.

- Inheritance patterns may, on first consideration, appear to deviate from Mendelian principles. Closer examination reveals informative Mendelian ratios.

- The Mendelian dihybrid F1 heterozygous cross produces four phenotypic offspring classes (Figure 2.8). When epistasis exists between two genes, three or two F2 phenotypes are observed.

- Importantly, the F2 phenotype ratios will clearly be a modification of the standard 9:3:3:1 ratio. The different F2 progeny phenotypes present as proportions of 16, indicating that two independently assorting genes are involved.

- Different F2 progeny segregation ratios indicate different kinds of epistatic interaction:

  ○ recessive epistasis (F2 ratio = 9:4:3)

  ○ dominant epistasis (F2 ratio = 12:3:1)

  ○ dominant suppression (F2 ratio = 13:3)

  ○ complementary gene action (F2 ratio = 9:7)

  ○ duplicate gene action (F2 ratio = 15:1).

Recessive epistasis

- **Recessive epistasis** occurs when expression of alleles at one locus are suppressed in individuals which are recessive homozygotes at a second 'epistatic' locus.

- This effect determines coat colour in Labrador and retriever dogs.

  ○ A dominant allele (**B**) at one locus results in black fur; the recessive homozygote (**bb**) is chocolate.

  ○ Recessive alleles at another locus (**E/e**) can prevent production of either black or chocolate fur. All **ee** individuals, regardless of the alleles at the **B/b** locus, are golden. There is no suppression when a dominant **E** allele is present.

- For example, Table 3.15 shows how the crossing of heterozygous F1 black-furred dogs (**BbEe**) produces dogs in the proportion of nine black-furred (**B-E-**):three chocolate-furred (**bbE-**):four golden-furred (**B-ee** and **bbee**).

- It is this modified F2 phenotype segregation ratio of 9:3:4 that indicates recessive epistasis.

F1 gametes	BE	Be	bE	be
BE	**BBEE** black	**BBEe** black	**BbEE** black	**BbEe** black
Be	**BBEe** black	**BBee** golden	**BbEe** black	**Bbee** golden
bE	**BbEE** black	**BbEe** black	**bbEE** chocolate	**bbEe** chocolate
be	**BbEe** black	**Bbee** golden	**bbEe** chocolate	**bbee** golden

Table 3.15 Epistatic expression of fur colour in the Labrador dog: crossing F1 heterozygotes

Epistasis

- The epistatic interaction controlling fur colour has a molecular explanation.
    - Allele **B** produces a functional enzyme required for the production of the dark pigment melanin.
    - **bb** individuals lack this enzyme so accumulate a lighter, chocolate pigment in their hairs.
    - The epistatic locus determines whether the pigment can enter the hair.
    - In **ee** (golden) individuals pigment deposition is largely prevented.

Dominant epistasis

- **Dominant epistasis** occurs when a dominant allele at the epistatic locus suppresses expression of another gene.
- Squash fruits can be white, yellow, or green. A dominant allele (**C**) at one locus results in yellow fruit, while recessive homozygotes (**cc**) are green. A dominant allele (**S**) at a second locus suppresses colour.
- Dominant epistasis can be detected by a modified F2 phenotypic ratio of 12:3:1.
- For example, Table 3.16 shows how crossing heterozygous F1 white squash-producing plants (**CcSs**) produces F2 progeny in the ratio 12 white-fruiter (C-S-/ccS-):3 yellow-fruiters (C-ss):1 green-fruiter (ccss).

Other epistatic interactions

- The shape of chicken combs (Table 3.17) results from an epistatic interaction between two genes, which does not change the standard F2 dihybrid ratios, as four phenotypic classes (i.e. different comb shapes) result.
- Interaction of the pea-comb allele (**P**) and the rose-comb allele (**R**) produce the comb shapes:
    - walnut (**P-R-**)
    - single (**pprr**)

F1 Gametes	CS	Cs	cS	cs
CS	CCSS white	CCSs white	CcSS white	CcSs white
Cs	CCSs white	CCss yellow	CcSs white	Ccss yellow
cS	CcSS white	CcSs white	ccSS white	ccSs white
cs	CcSs white	Ccss yellow	ccSs white	ccss green

Table 3.16 Epistatic expression of squash fruit colour: crossing F1 heterozygotes

F1 Gametes	PR	pR	Pr	pr
PR	PPRR walnut	PpRR walnut	PPRr walnut	PpRr walnut
pR	PpRR walnut	ppRR rose	PpRr walnut	ppRr rose
Pr	PPRr walnut	PpRr walnut	PPrr pea	Pprr pea
pr	PpRr walnut	ppRr rose	Pprr pea	pprr single

Table 3.17 Comb shape among F2 progeny from crossing F1 heterozygous walnut-comb chickens (PpRr)

○ pea (**P-rr**)

○ rose (**ppR-**).

• An F1 cross of two double heterozygous walnut (PpRr)-combed fowl produces offspring in the ratio of 9 walnut:3 pea:3 rose:1 single (Table 3.17).

Complementary gene action

• **Complementary genes** work sequentially to produce full phenotypic expression. Considering two loci:

  ○ the full phenotype is only expressed when both loci have at least one dominant allele

  ○ homozygosity for recessive alleles at either or both of the two loci causes an alternative non-expressing phenotype

  ○ complementary gene action can be recognized when the 9:3:3:1 F2 phenotypic ratio is modified to a 9:7 expressing/non-expressing phenotypic ratio.

• For example, when two pure-breeding white-flowering strains of the sweet pea *Lathyrus odorata* are crossed, F1 purple flowering plants are produced. When these are interbred, F2 progeny are produced in the ratio of 9 purple to 7 white flowering plants (Figure 3.5).

• Figure 3.5 shows how a dominant allele must be present at each of two loci (**C/c** and **P/p**) to produce coloured sweet pea petals.

(A)

Parents	White flowering **ccPP**	White flowering **CCpp**
Gametes	**cP**	**Cp**
F1	**CcPp** Purple flowering	

(B)

F1 Gametes	CP	Cp	cP	cp
CP	**CCPP** purple	**CCPp** purple	**CcPP** purple	**CcPp** purple
Cp	**CCPp** purple	**CCpp** white	**CcPp** purple	**Ccpp** white
cP	**CcPP** purple	**CcPp** purple	**ccPP** white	**ccPp** white
cp	**CcPp** purple	**Ccpp** white	**ccPp** white	**ccpp** white

Figure 3.5 The inheritance of petal colour in sweet peas. (A) Production of F1 heterozygous purple-petalled plants (CcPp) from two white-flowering strains. (B) Production of the F2 generation.

Epistasis

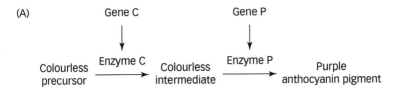

(A)

Gene C   Gene P

Colourless precursor → Enzyme C → Colourless intermediate → Enzyme P → Purple anthocyanin pigment

(B)

Genotype	Functional enzyme present	Sweet pea petal colour
C–P–	Enzymes C and P	Purple; anthocyanin produced
C–pp	Enzyme C	White; no anthocyanin produced
ccP–	Enzyme P	White; no anthocyanin produced
ccpp	None	White; no anthocyanin produced

Figure 3.6 The genetic control of pigment production in sweet peas. (A) Enzyme products of genes C and P catalyse pigment production. (B) Relationship between genotype, functional enzyme, and flower colour.

- There is a biochemical explanation for the determination of sweet pea petal colour. The two genes produce enzymes involved in a biochemical pathway of anthocyanin pigment synthesis (Figure 3.6).
- Only those plants with active forms of both enzymes C and P (individuals with genotype **C-P-**) will be purple. All other genotypes lack at least one enzyme and so result in white flowers.

Duplicating genes

- Complementary gene products work sequentially. Occasionally, a situation is identified in which gene products work in parallel, i.e. there are two (**duplicate**) ways of achieving the same phenotype.
- Duplicating genes can be recognized when the 9:3:3:1 F2 phenotypic ratio is modified to 15:1. Only the double recessive homozygote shows the alternative (non-expressing) phenotype.
- Wheat kernels are generally coloured. Two different enzymes (A or B) can convert the colourless precursor molecule into the red (anthocyanin) pigment. Only when both enzymes A and B are non-functional is a white kernel produced (Figure 3.7).

Check your understanding

3.7 What genetic crosses would you do, and what results would you expect, in order to work out whether two loci were involved in complementary gene action or recessive epistasis?

(A)

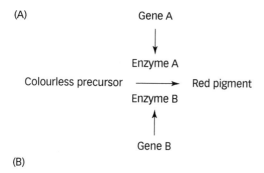

(B)

Genotype	Kernel phenotype	Enzymatic activity
9 A–B–	Coloured	Functional enzyme from both loci
3 aaB–	Coloured	Functional enzyme from A locus
3 A–bb	Coloured	Functional enzyme from B locus
1 aabb	White	No functional enzyme

Figure 3.7 The genetic control of wheat kernel production. (A) Both enzyme A and enzyme B catalyse pigment production. (B) Crossing F1 heterozygotes (AaBb) produces an F2 ratio of 15 coloured to 1 colourless kernelled plants.

Looking for extra marks?

A contrasting genetic situation to epistasis, where several genes influence expression of a single trait, is **pleiotropy**, whereby a single gene is responsible for several different phenotypic effects. For example, the accumulation of phenylalanine, which results from a defective PKU gene (and faulty phenylalanine hydrolase), results in a 'mousy odour', difficulty in walking, broad shoulders, light skin, and mental retardation.

3.7 PEDIGREE ANALYSIS

- Pedigree analysis is the working tool of human genetics.
- A **pedigree** (or family tree) illustrates the relationships between members of a family.
- A pedigree shows which individuals express a given trait.
- The expression pattern of a trait through several generations gives clues to its mode of inheritance, e.g. dominant or recessive; autosomal, or X-linked.
- A human pedigree is often constructed in response to an individual presenting with a medical problem. This individual, termed the **proband**, is identified in a pedigree with an arrow.

Pedigree analysis

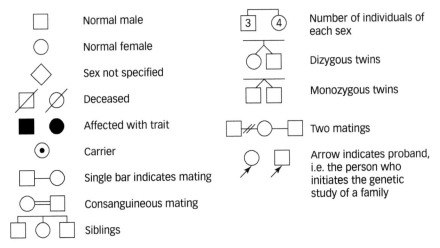

Figure 3.8 Standard pedigree symbols.

- There is a universally-agreed pedigree notation (Figure 3.8).
- Table 3.18 presents idealized pedigrees for the inheritance of autosomal and X-linked single gene conditions.
- Attention is drawn to key features of the example pedigrees that aid identification of the genetic basis of an inherited condition.
- Various genetic phenomena complicate the interpretation of pedigree inheritance patterns. These compounding factors are discussed in the following sections and include:
  - reduced penetrance
  - age-dependent penetrance
  - new mutations
  - locus heterogeneity
  - mitochondrial inheritance
  - **genomic imprinting**.

Check your understanding

3.8 A couple each have a sister with cystic fibrosis. They want to know the probability that they might have a child with cystic fibrosis. No other member of the couple's immediate family suffers from cystic fibrosis. (NB: first, draw their pedigree.)

Reduced penetrance

- An individual may have a disease-expressing genotype, but not express the associated phenotype.

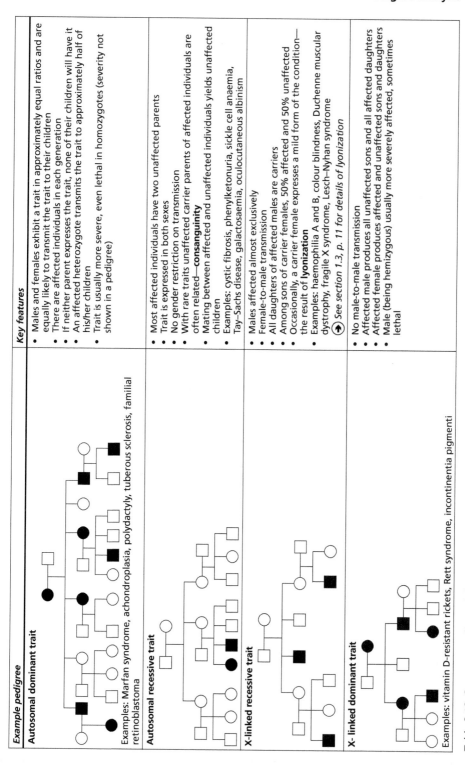

Example pedigree	Key features
Autosomal dominant trait Examples: Marfan syndrome, achondroplasia, polydactyly, tuberous sclerosis, familial retinoblastoma	• Males and females exhibit a trait in approximately equal ratios and are equally likely to transmit the trait to their children • There are affected individuals in each generation • If neither parent expresses the trait, none of their children will have it • An affected heterozygote transmits the trait to approximately half of his/her children • Trait is usually more severe, even lethal in homozygotes (severity not shown in a pedigree)
Autosomal recessive trait	• Most affected individuals have two unaffected parents • Trait is expressed in both sexes • No gender restriction on transmission • With rare traits unaffected carrier parents of affected individuals are often related—**consanguinity** • Mating between affected and unaffected individuals yields unaffected children • Examples: cystic fibrosis, phenylketonuria, sickle cell anaemia, Tay-Sachs disease, galactosaemia, oculocutaneous albinism
X-linked recessive trait	• Males affected almost exclusively • Female-to-male transmission • All daughters of affected males are carriers • Among sons of carrier females, 50% affected and 50% unaffected • Occasionally, a carrier female expresses a mild form of the condition—the result of **lyonization** • Examples: haemophilia A and B, colour blindness, Duchenne muscular dystrophy, fragile X syndrome, Lesch-Nyhan syndrome ⊕ See section 1.3, p. 11 for details of lyonization
X- linked dominant trait Examples: vitamin D-resistant rickets, Rett syndrome, incontinentia pigmenti	• No male-to-male transmission • Affected male produces all unaffected sons and all affected daughters • Affected female produces affected and unaffected sons and daughters • Male (being hemizygous) usually more severely affected, sometimes lethal

Table 3.18 Common pedigree inheritance patterns (human genetic disease examples)

Pedigree analysis

- **Penetrance** is the proportion of individuals that express the phenotype normally associated with the expressing genotype.
- The variable penetrance of an allele can be explained by a number of factors, e.g. by **allelic heterogeneity** when different alleles result in varied expression of a trait or by epistasis.
  ➔ *See section 3.6 (p. 47) for details of epistatic interactions.*
- Reduced (or incomplete) penetrance is detected in a pedigree when the expression pattern suggests an autosomal dominant trait, but there is an apparent anomaly, i.e. a non-expressing individual has a parent with the expressing dominant allele *and* appears to transmit the disease allele to the next generation (Figure 3.9).
- An **obligate carrier** is an individual known to possess a disease-causing allele, but who may or may not express the phenotype. Individual II-4 in Figure 3.9 is an obligate carrier.
- Penetrance (expression) rates for different conditions are estimated by examining a large number of families and determining the percentage of obligate carriers that express a disease.
- For retinoblastoma, 10% of obligate carriers do not express the disease, thus penetrance is 90%.
- Even if a genotype is always expressed phenotypically, there can be variation in the extent. **Expressivity** measures the degree of phenotypic expression and is influenced by epistatic and environmental factors.

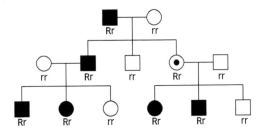

Figure 3.9 A pedigree illustrating the inheritance pattern of retinoblastoma (a malignant eye tumour) caused by a dominant allele, R, in the retinoblastoma gene *RB1*.

Age-dependent penetrance

- A delay in the age of onset of an inherited condition is termed **age-dependent penetrance**, i.e. a person has the disease-causing allele, but it is not expressed until later in life (Table 3.19).
- Age-dependent penetrance complicates the deduction of the mode of inheritance for a disease because it is not possible to know until later in life whether an individual carries the disease-causing mutation.
- The best known example is Huntington disease (characterized by uncontrolled limb movement and progressive dementia).
- An affected individual often has children before he/she is aware that they possess the disease-causing allele.

Diseases showing age-dependent penetrance	Brief description
Huntington disease	Neurodegenerative disease. Affects muscle coordination and cognitive functions
Haemochromatosis	Recessive disorder of liver iron storage
Adult polycystic kidney disease	Multiple cysts in both kidneys, which enlarge. Renal functions disrupted
Familial Alzheimer's disease (AD)	Development of ~10% AD promoted by dominant susceptibility alleles of genes *APP*, *PS1*, *PS2* and *APOE*
Breast cancer	*BRCA 1* and *2* mutations produce a 60–80% lifetime risk of developing breast cancer

Table 3.19 Examples of human diseases showing age-dependent penetrance

- Delayed age of onset of a disease reduces **natural selection** against a disease-causing allele, thus increasing its frequency in a population.
- **Anticipation** is often associated with age-dependent penetrance, i.e. expression of the condition at earlier ages and with greater severity in succeeding generations.

New mutation

- The occurrence of a new mutation is always a possible explanation for a single individual in a pedigree exhibiting a trait.
- A new mutation is likely to have occurred in the relevant gene in a parent's germ cell.
- There is no elevated recurrence risk for subsequent children (unless the new mutation is a germ-line mutation).
- Care must be taken when interpreting such pedigrees to eliminate the possibility of mild manifestations of a condition in other family individuals, e.g. expression of neurofibromatosis 1 (NF1) is highly variable.
- Careful analysis is needed to distinguish an apparent new mutation from the appearance of a homozygous recessive individual produced by carrier parents.
- The appearance of new mutations in a gene is common for some dominant disorders, e.g. achondroplasia, NF1, and tuberous sclerosis.

Locus heterogeneity

- The pedigree produced from one family may suggest a condition has an autosomal dominant mode of inheritance, while another family's pedigree suggests a recessive pattern.
- This apparent contradiction is commonly caused by **locus heterogeneity**, i.e. whereby mutations in different genes (in different families) cause the same phenotype (Table 3.20).

Mitochondrial inheritance

- If a disease is expressed in both sexes, but transmitted from the female parent (i.e. maternally inherited) it is likely to be caused by a mutation in a mitochondrial gene (Table 3.21).

Pedigree analysis

Disease	Description	Chromosomal locations (gene)
Osteogenesis imperfecta	Brittle bone disease; bones fragile and prone to fracture	ch17 (*COL1A1*) ch7 (*COL1A2*)
Adult polycystic disease	Accumulation of renal cysts leading to kidney failure	ch16 (*PKD1*) ch4 (*PKD2*)
Tuberous sclerosis	Non-malignant tumours in brain and other organs; seizures, developmental delay, behaviour problems	ch9 (*TSC1*) ch16 (*TSC2*)
Familial melanoma	Development of melanomas	ch1 (*CMM1*); ch9 (*CMM2*)

Table 3.20 Examples of human diseases showing locus heterogeneity

- All offspring of diseased mothers show the trait, while none of the offspring of diseased fathers express disease phenotype (Figure 3.10).
  - ➔ *See section 3.5 (p. 45) for details of maternal inheritance.*
- Mitochondria have their own DNA molecules encoding a small number of genes.
  - ➔ *See section 6.7 (p. 120) for details of organelle genetic systems.*
- Mixed populations of mutated and normal mitochondrial DNA, termed **heteroplasmy**, often occur in cells.
- Depending upon the proportion of mutated and normal genomes in relevant tissue cells, mitochondrial conditions often show:
  - variability of expression between different individuals
  - reduced penetrance.
- Complete penetrance of the disease causing mutation is shown in Figure 3.10.

Genomic imprinting

- Various genes in mammals, flowering plants, and insects are expressed differently depending on the parental origin of an allele.
- Either the maternal or paternal allele is transcriptionally inactive, i.e. it is **imprinted**:
  - **paternal imprinting**—silenced paternal allele, expressed maternal allele
  - **maternal imprinting**—silenced maternal allele, expressed paternal allele.

Mitochondrial disease	Brief description	Gene mutated
Ragged-red fibre syndrome (MERF)	Epilepsy, dementia, ataxia, myopathy	Caused by mutations in NADH dehydrogenase subunit 5 or one of 5 mt tRNA genes
Mitochondrial encephalomyopathy and stroke-like episodes (MELAS)	Seizures, myopathy, deafness, dementia	Caused by mutations in NADH dehydrogenase subunit 5 or one of 3 mt tRNA genes
Leber hereditary optic neuropathy (LHON)	Rapid vision loss due to optic nerve death	NADH dehydrogenase subunit 4 (G to A substitution)
Kearns–Sayre disease	Muscle weakness, cerebellar damage, heart failure	Variable (1.1–10kb) deletion of mtDNA

Table 3.21 Examples of human mitochondrial (mt) diseases

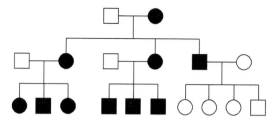

Figure 3.10 A pedigree illustrating a typical inheritance pattern for a mitochondrial-encoded condition.

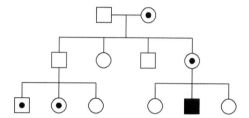

Figure 3.11 A pedigree illustrating inheritance of the deletion that produces Prader–Willi syndrome. The deletion has been silently passed through I-1, II-4 to III-5, who expresses the condition.

Tip: for any pedigree showing the inheritance of an imprinted allele it helps to work out each genotype by:
  (i)   representing the relevant pair of homologous chromosomes by two vertical lines
  (ii) indicating on the linear chromosome whether there is an active, silenced, or deleted allele.

- Allele silencing is achieved by heavy methylation of the DNA, plus hypoacetylation of associated histones and chromatin condensation.
- As normal individuals have only one transcriptionally active allele, if this is mutated a changed phenotype (disease) results.
- Initial pedigree analysis of the inheritance of an imprinted gene often suggests a recessively inherited trait (Figure 3.11).
- Clues to genomic imprinting come from the pleiotropic nature of an imprinted condition (Table 3.22).

Condition	Nature of alleles	Symptoms
Prader–Willi syndrome	Silenced maternal allele Mutated paternal allele	Learning difficulties, short stature, compulsive eating
Angelman syndrome	Silenced paternal allele Mutated maternal allele	Learning difficulties, speech problems, jerky movements, unusually happy disposition

Table 3.22 Two imprinted human conditions: Prader–Willi and Angelman syndromes, which result from a microdeletion of chromosome 15q11-q13

Looking for extra marks?

The **genetic conflict hypothesis** suggests that genomic imprinting evolved as the result of differing interests for each parent in terms of the evolutionary **fitness** of certain genes involved in foetal/neonate development.

Paternal interest was best served by growth of large offspring at the expense of the mother.

Maternal interest was to conserve resources for her own survival, while providing nourishment for her litter.

In support, paternal expressed genes tend to be growth promoters, while maternal expressed genes limit growth.

Check your understanding

3.9 What is the difference between (i) multiple allelism and locus heterogeneity, and (ii) sex-linked and genomic imprinted traits?

3.10 In humans, how would you distinguish between an autosomal dominant and a mitochondrial condition? What features of autosomal dominance and mitochondrial inheritance can make them difficult to identify?

3.11 Explain why mutations in transcription factors, such as the developmental transcription factor TBX5, are generally pleiotropic.

online resource centre

You'll find guidance on answering the questions posed in this chapter—plus additional multiple-choice questions—in the Online Resource Centre accompanying this revision guide. Go to **www. oxfordtextbooks.co.uk/orc/thrive/** or scan this image:

4 Eukaryotic Gene Mapping

This uses a variety of genetic and molecular techniques to locate precisely the chromosomal position of genes.

Key concepts

- Eukaryotic gene mapping is often a two-part process involving:
  - (i) **genetic mapping**, which uses traditional Mendelian analysis of carefully constructed genetic crosses to produce maps showing the *relative* positions of genes and other features on a chromosome; this analysis identifies *approximate* gene positions
  - (ii) **physical mapping** employs molecular techniques, e.g. DNA sequencing, to directly examine the DNA of chromosomes to determine the *precise* position of genes.
- Genetic mapping analyses the inheritance patterns of linked genes.
- Genetic maps are constructed from the results of two- and three-point test crosses
- The number of recombinant progeny are used to determine the relative order of genes and distances between them.
- Human genes are mapped by examining pedigrees for the co-segregation of traits
- Logarithm of odds (LOD) scores are calculated, using data from human pedigrees, to assess the likelihood of linkage of two genes.

Linkage and recombination

Genetic mapping requires a good understanding of the chromosomal events of meiosis and to be able to visualize relationships between alleles of genes and their location on chromosomes.

➡ *See section 1.6 (p. 16) for details of meiosis.*

4.1 LINKAGE AND RECOMBINATION

- **Linked genes** are located on the same chromosome.
- Alleles of linked genes are inherited together unless crossing over of non-sister chromatids occurs between the two genes during prophase 1 of meiosis.
- Crossing over results in **recombination**, i.e. new allele combinations.
- Gametes with the new combinations of alleles are termed **recombinant** gametes.
- Consider meiosis in an individual heterozygous for two linked genes, **A/a** and **B/b** (Figure 4.1):
  - crossing over occurs during prophase I (4.1b)
  - this produces two recombinant chromatids (4.1c)
  - at the end of meiosis II, two gametes contain parental-type chromosomes and two recombinant chromosomes (4.1d).

As an aid to understanding the behaviour of linked alleles during meiosis it is helpful to indicate genotypes by positioning allele letters alongside lines drawn to represent chromosomes (as in Figure 4.1).

- In the one meiosis represented in Figure 4.1 there is an equal number of parental and recombinant gametes.
- Overall, during the many meioses that produce gametes for sexual reproduction, recombinant gametes are in the minority: they are only produced if crossing over 'happens' to occur between two linked genes.
  - Most of the gametes will have the same allele combinations as the two parental chromosomes (A B and a b).

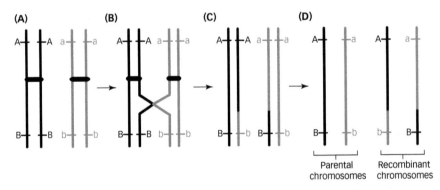

Figure 4.1 The production of recombinant gametes in an individual heterozygous for genes A and B. Note that the four chromosomes shown on the far right (two parental and two recombinant) will enter separate gametes.

- ○ There will generally be only a few recombinant gametes, with a changed allele combination (A̲ b̲ and a̲ B̲).
- The number of recombinant gametes depends on how far apart two genes are on a chromosome.
- The further apart two genes are, the greater the probability that a crossover event will occur between them and so the greater the number of recombinant gametes.
- Consider three linked genes: *M/m*, *N/n*, and *Q/q*:

M	N	Q
m	n	q

- A crossover will occur in more meioses between genes *M* and *N*, than between genes *N* and *Q*. The frequency of recombinant gametes M/n and m/N will thus be greater than the frequency of recombinant gametes N/q and n/Q.
- If two genes are located at opposite ends of chromosomes, a crossover is likely to occur between them during every meiosis, so 50% of gametes will be recombinants, i.e. the four different types of gametes (two parental and two recombinant are produced in equal proportion).
- Likewise, when two genes are located on different pairs of homologous chromosomes, owing to independent assortment of their alleles during meiosis, gametes of four different genotypes in equal proportions are produced (Figure 4.2)
  ➔ *See section 2.3 (p. 30) for details of independent assortment of alleles.*

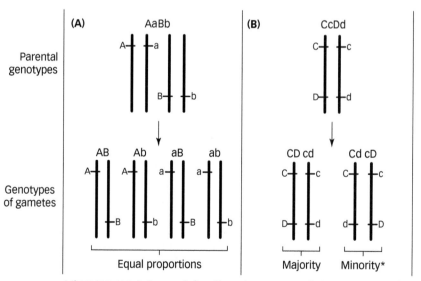

* these two gamete types only form if crossing over occurs between genes C and D

Figure 4.2 Gamete formation by double heterozygotes. (A) Heterozygote AaBb (genes *A* and *B* are on different chromosomes). (B) Heterozygote CcDd (genes *C* and *D* are linked).

4.1 What affects the chances of recombination between two loci on the same chromosome?

4.2 GENE MAPPING USING DIHYBRID CROSSES

- The standard gene mapping cross is between a double heterozygous individual (AaBb) and another who is homozygous recessive (aabb) for both genes.
- Each gene encodes a different trait.
- The proportions of the different types of progeny indicate:
  - (i) whether two genes are linked or on different chromosomes
  - (ii) if linked, how close together they are.
- If the two genes in a dihybrid mapping cross are on different chromosomes (are independently assorting) there will be four different progeny phenotypes in equal proportions (Figure 4.3).
- If the two genes under investigation are linked, a double heterozygous×double homozygous cross produces major deviations from the 1:1:1:1 phenotypic ratio (Figure 4.4).
- Generally, four different progeny phenotypes are obtained, but the two **parental** phenotypes greatly outnumber the two recombinant phenotypes.
- Phenotype ratios, such as 28:1:1:28 or 74:1:1:74, result (Figure 4.4).

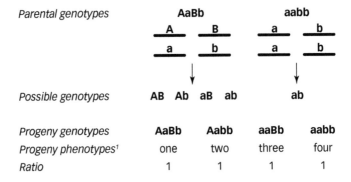

¹Phenotype 1: trait A and trait B dominantly expressed
Phenotype 2: trait A dominantly expressed, and trait B recessively expressed
Phenotype 3: trait A recessively expressed, and trait B dominantly expressed
Phenotype 4: trait A and trait B recessively expressed

Figure 4.3 Progeny phenotype ratios resulting from mating double heterozygous and double homozygous recessive individuals when genes A and B are on different chromosomes.

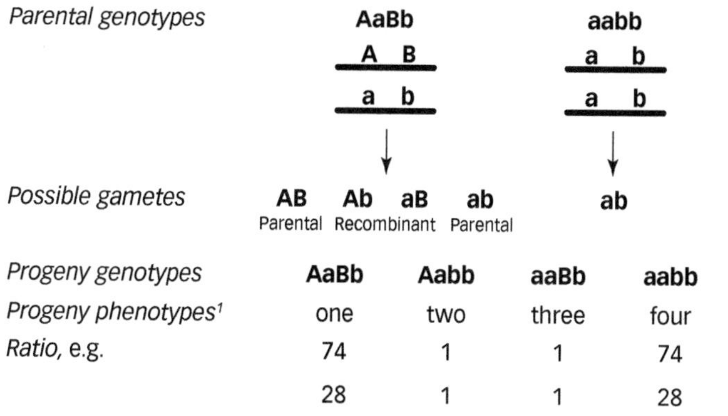

Parental genotypes	**AaBb**				**aabb**
	A B				a b
	a b				a b
	↓				↓
Possible gametes	**AB**	**Ab**	**aB**	**ab**	**ab**
	Parental	Recombinant	Parental		
Progeny genotypes	**AaBb**	**Aabb**	**aaBb**	**aabb**	
Progeny phenotypes[1]	one	two	three	four	
Ratio, e.g.	74	1	1	74	
	28	1	1	28	

[1]Phenotype 1: trait A and trait B dominantly expressed
Phenotype 2: trait A dominantly expressed, and trait B recessively expressed trait 2
Phenotype 3: trait A recessively expressed, and trait B dominantly expressed
Phenotype 4: trait A and trait B recessively expressed

Figure 4.4 Progeny phenotype ratios resulting from mating double heterozygous and double homozygous recessive individuals when genes A and B are linked.

- A deviation from the standard 1:1:1:1 phenotype ratio can be confirmed statistically by a chi-squared (χ^2) test.
  ➡ *See section 2.6 (p. 34) for details of using a χ^2 test.*
- In mapping crosses the size of the deviation from the 'unlinked' 1:1:1:1 progeny phenotype ratio indicates the distance between two genes: the greater the deviation, the closer together are the two genes under investigation.
- Considering the two hypothetical ratios of 28:1:1:28 or 74:1:1:74 of Figure 4.4, the latter ratio indicates closer genes, i.e. there was less opportunity for crossing over to generate recombinant gametes and so recombinant offspring.
- Occasionally, a mapping cross produces progeny showing just two different phenotypes in a 1:1 ratio, which indicates that the two genes under study are so close together that no crossover occurred between the two genes in any meiosis.
- Conversely, a 1:1:1:1 progeny phenotype ratio would be obtained if two genes are sufficiently distant on the same chromosome that a chiasma forms between them during every meiosis (Table 4.1).

Calculating the genetic distance between two linked genes

- If the inheritance pattern of two traits in a dihybrid mapping cross shows the two encoding genes to be linked (i.e. there is a deviation from a 1:1:1:1 progeny phenotypic ratio) the results can also be used to estimate the **genetic distance** between the two genes (see Box 4.1).

$$\text{The \% of recombinants} = \frac{\text{number of recombinants}}{\text{total number of progeny}}$$

Gene mapping using dihybrid crosses

Genetic situation	Progeny genotypes	Genotype proportions (%)
Genes on different chromosomes or widely separated on the same chromosome	AaBb parental	25
	Aabb recombinant	25
	aaBb recombinant	25
	aabb parental	25
Linked genes: chiasmata form between the two genes during some meioses	AaBb parental	>25
	Aabb recombinant	<25
	aaBb recombinant	<25
	aabb parental	>25
Tight linkage: genes close together with no crossover events between them	AaBb parental	50
	Aabb parental	50

Table 4.1 Progeny genotypes resulting from a dihybrid mapping cross: AaBb×aabb

- The genetic distance is expressed in **map units** or **centimorgans** (**cM**).
- **1 cM** is the distance between two genes that produces 1% recombination.
- 1 cM is approximately equivalent to 10^6 nucleotides. This value is approximate as the frequency of crossovers varies in different regions of a chromosome.

Box 4.1 Determining the genetic distance between two Drosophila genes

- In *Drosophila melanogaster* vestigial wings (**vg**) are recessive to normal wings (**vg⁺**), and purple eyes (**pr**) are recessive to red eyes (**pr⁺**).
- In 1910, Thomas Morgan performed a cross between heterozygous normal-winged, red-eyed flies (**pr⁺ pr vg⁺vg**) and homozygous recessive vestigial-winged, white-eyed flies (**pr pr vg vg**). The following results were obtained:

Progeny (genotype/phenotype)		Progeny number	
pr⁺pr vg⁺vg	red-eyed normal-winged	1339	Parental
pr⁺pr vg vg	red-eyed vestigial-winged	151	Recombinant
pr pr vg⁺vg	purple-eyed normal-winged	154	Recombinant
pr pr vg vg	purple-eyed vestigial-winged	1195	Parental

- The progeny phenotype ratio is approximately 9:1:1:9. This is a clear deviation in progeny proportions from the 1:1:1:1 phenotype ratio gained when two genes are located on separate chromosomes.
- Thus, this 9:1:1:9 phenotype progeny ratio indicates that the genes for eye colour and wing type are linked.
- The genetic distance between the two genes is calculated from the percentage of recombinants:

$$\% \, \text{Recombinants} = \frac{\text{number of recombinants}}{\text{total number of progeny}} = \frac{305}{2839} = 0.107 \text{ or } 10.7\%$$

- A recombination frequency of 10.7% indicates a genetic distance of 10.7 cM, i.e. the gene determining red/purple eyes is 10.7 cM from the gene controlling wing production.

Coupling and repulsion

- Consider a heterozygote **CcDd**; genes *C/c* and *D/d* are linked.
- If both dominant alleles are on one chromosome, and both recessive alleles are located on the other, this is referred to as the **coupled**, or **cis configuration**.
- The **linkage** of a dominant allele of one gene and the recessive allele of the other is termed the **repulsion** or **trans configuration** (Table 4.2).
- It is important to know whether two pairs of alleles are coupled or in repulsion, as the arrangement affects the results of a test cross.
- Test cross progeny phenotypes are the same, but their relative numbers differ depending on whether alleles are in coupling or in repulsion.
- The more numerous progeny types always represent the parental ones, even if it seems counterintuitive! Study Figure 4.5.

Coupled		*Repulsion*	
C	D	C	d
c	d	c	D

Table 4.2 Alternative allele arrangements of linked genes

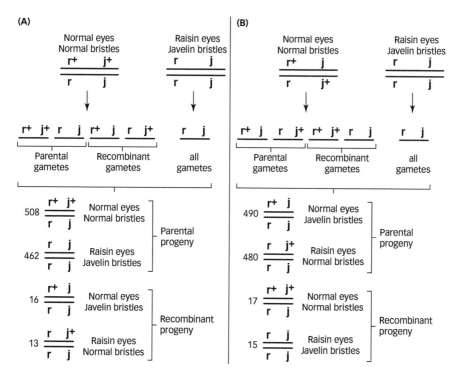

Figure 4.5 The results of two test crosses. (A) Alleles are in coupling arrangement. (B) Alleles are in repulsion arrangement (genes *r⁺/r* and *j⁺/j* determine raisin eyes and javelin bristles in *Drosophila melanogaster*).

> ### Check your understanding
>
> 4.2 What kind of gametes, and in what proportions, may be made by an individual who develops from fertilization of an egg carrying alleles **m** and **n** by a sperm carrying alleles **M** and **N**, if the two gene loci are (i) on different pairs of homologous chromosomes; (ii) adjacent to each other on the same pair of homologous chromosomes; (iii) 7 cM apart on the same pair of homologous chromosomes?

4.3 CONSTRUCTING A GENETIC MAP

- A genetic map is a linear representation of a chromosome showing the relative distances between genes and other chromosomal landmarks.
- Genetic maps are produced from dihybrid mapping crosses in the following way:
  - (i) a series of pair-wise mapping crosses (double heterozygotes with corresponding double recessive homozygotes) are made involving the relevant genes
  - (ii) genetic distances (in cM) are calculated from recombination frequencies between each pair of genes
  - (iii) a gene order is deduced from the various genetic distances by a trial and error process.
- Consider mapping genes *A*, *B*, *C* and *D* (Table 4.3).
- The mapping crosses established a relative gene order, *but* there are two things to note.
  - (i) It is impossible from the mapping cross data to know whether genes *B*, *C*, and *D* should be positioned to the left or to the right of gene *A* on the genetic map.
  - (ii) The genetic map is not additive. The distance calculated from recombinants produced in the cross involving genes *A* and *D* suggests these two genes are

Genes involved in mapping cross	Distance (cM) between genes
A and B	3.2
B and C	6.8
C and D	2.0
B and D	7.5
A and D	10.2

Table 4.3 Genetic distances between genes A, B, C, and D. These were calculated from recombination frequencies among progeny of mapping crosses that involved pairwise combinations of the four genes

10.2 cM apart. Yet, if we sum the distances calculated from the separate pairwise crosses between genes *A* and *B*, *B* and *C*, and *C* and *D*, the distance is 12 cM.

- Whether genes *B*, *C*, and *D* are located to the left or right of gene *A* can be established by test-crossing with other genes or a physical landmark, such as a centromere, whose precise chromosomal position is already known.
- The discrepancy in genetic distance between genes *A* and *D* is caused by **double crossovers**. This inconsistency in genetic distance can be resolved by conducting **trihybrid mapping crosses** (section 4.4).

Double crossovers

- If two genes are far apart on a chromosome two crossover events (a double crossover) may occur between them.
- The construction of genetic maps relies on the occurrence of a single crossover event between two genes to generate recombinant progeny whose numbers are used to calculate recombination frequencies and so map distances.
- As the consequence of a double crossover, alleles remain in their original relationships. No recombinant gametes, and therefore recombinant offspring, are produced (Figure 4.6).

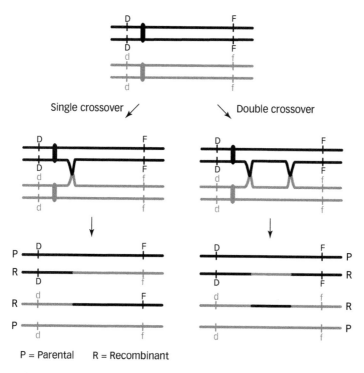

P = Parental   R = Recombinant

Figure 4.6 The consequences of a single or double crossover between a pair of linked genes. The nature of the resultant chromosomes (and so gametes) are shown.

Constructing a genetic map

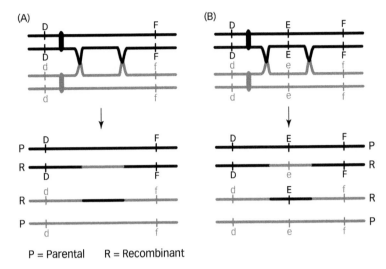

P = Parental   R = Recombinant

Figure 4.7 The consequences of a double crossover between (A) a pair of linked genes and (B) three linked genes. The nature of the resultant chromosomes (and so gametes) is shown.

- It is impossible in a dihybrid mapping cross to distinguish between the progeny produced by two crossover events and the absence of any crossovers.
- The consequence of a double crossover is that genetic distances between genes are underestimated; genetic maps constructed from short distances are usually more accurate than those based on longer distances.
- A double crossover can, however, be detected if a third gene, located between the two genes being mapped, is included in a mapping cross. Recombinant chromosomes with different allele combinations are produced (Figure 4.7).
- Figure 4.7B illustrates the value of three-point crosses in identifying double crossovers. The alleles of the outer two genes (genes *D* and *F*) are in the parental (apparently non-recombinant) arrangement, but the alleles of the middle gene (*E*) are arranged differently.
- In total, six different recombinant chromosomes (and so gametes) can be produced as the result of two different single crossover events or a double crossover between three linked genes (Figure 4.8).

Check your understanding

4.3 Consider the following pair of homologous chromosomes and crossovers occurring in regions 1 and 2 between allele pairs P/p, Q/q and R/r.

P   q   r

p   q   R

    Region 1   Region 2

Would the chromosomes produced as the result of crossovers be informative for linkage mapping if the crossovers occurred in (i) region 1 only; (ii) region 2 only; (iii) in both regions simultaneously?

4.4 GENE MAPPING USING TRIHYBRID CROSSES

- A trihybrid mapping cross involves mating an individual heterozygous at three loci (**DdEeFf**) with another homozygous recessive for these three genes (**ddeeff**).
- Compared with a dihybrid mapping cross, it:
  - yields more accurate information, as the progeny from double crossovers can be identified
  - is more efficient because the order of three genes can be established from analysing the nature of the progeny of just one cross.
- In a trihybrid mapping cross, when individual DdEeFf is crossed with a triple recessive homozygote (ddeeff), eight classes of progeny are produced, which fall into four categories:
  - parental (non-recombinant) progeny

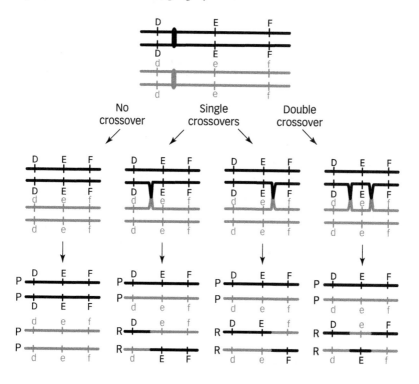

P = Parental   R = Recombinant

Figure 4.8 Parental and recombinant gametes produced by single and double crossovers between three linked genes.

Gene mapping using trihybrid crosses

- o double crossover recombinant progeny
  - o recombinant progeny resulting from a single crossover between genes D and E
  - o recombinant progeny resulting from a single crossover between genes E and F.
- Relate the above progeny classes to the gametes of Figure 4.8. Each of the eight different types of gametes (two parental and six recombinant) fuses with one which contains the chromosome d e f from the ddeeff individual.
- There is a standard way of identifying the different progeny classes of a trihybrid cross which then:
  - o yields information on the order of the three genes
  - o enables distances between genes to be calculated.
- The procedure for analysing a trihybrid cross is best considered in the context of a worked example.

An example: determining genetic distances between three tomato genes

- Tomatoes heterozygous for fruit shape (peach/normal), height (dwarf/normal), and leaf mottling (present/absent) were crossed with triple recessive homozygotes.
- Progeny phenotypes and numbers are presented in Table 4.4.
- First, establish **gene order**.
  - o Decide which classes are the parental progeny (i.e. those progeny produced when *no* crossing over occurred). These are the two classes with the most progeny (Table 4.4, rows 1 and 2).
  - o Identify the double crossover classes of progeny. These are the two classes with the smallest number of progeny (Table 4.4, rows 7 and 8).
  - o Compare the phenotypes of the parental and double crossover progeny. They should be alike in two characteristics and differ in one.
  - o The characteristic that differs between the double crossover and parental progeny is encoded by the middle gene. It will have been this gene's alleles that changed positions as the result of the double crossover.
  - o In the tomato example (Table 4.4, classes 1, 2, 7, and 8), peach and mottled recessive alleles have stayed separate. It is the dwarf allele that has swapped chromosomal position. The dwarf (D/d) locus is in the middle.
  - o There is now a gene order: peach dwarf mottled (or mottled, dwarf, and peach—further crosses are necessary to work out which way around it might be).
  - o Also note that the repulsion arrangement of alleles on the parental chromosomes (Table 4.4, column 3, rows 1 and 2): one recessive allele is located on one homologue and the other two recessive alleles on the other.
- Having worked out a gene order, **gene distances** can then be calculated from the recombination frequencies between each pair of genes, i.e. between the peach and dwarf genes, and between the dwarf and mottled genes.

Progeny phenotype	Progeny numbers	Progeny genotype	Nature of progeny
Peach	402	pDM/pdm	Parental
Dwarf, mottled	444	Pdm/pdm	
Dwarf	62	PdM/pdm	Recombinant (single crossover between genes)
Mottled, peach	54	pDm/pdm	
Wild type	17	PDM/pdm	Recombinant (single crossover between genes)
Dwarf, mottled, peach	19	pdm/pdm	
Dwarf, peach	1	pdM/pdm	Recombinant (double crossover)
Mottled	2	PDm/pdm	

Table 4.4 The results of a trihybrid tomato mapping cross between parents of genotypes **pDM/Pdm** and **pdm/pdm***

*pDM/Pdm is an alternative representation of the heterozygous parent PpDdMm and pdm/pdm of the homozygous recessive parent, ppddmm.
Allele symbols: D/d = normal height/dwarf; P/p = normal fruit shape/peach; M/m = normal leaves/mottled.

- In each case the number of recombinants includes the progeny from both a single crossover and the double crossover (Table 4.4).

$$\% \text{ recombinants between peach and dwarf} = \frac{17+19+1+2}{1002} = 0.039 \text{ or } 3.9\%$$

$$\% \text{ recombinants between mottled and dwarf} = \frac{62+54+1+3}{1002} = 0.12 \text{ or } 12\%$$

- Thus, gene distances between:

peach and dwarf = 3.9 cM

mottled and dwarf = 12 cM.

- As aforementioned there are two alternative arrangements:

Peach   dwarf   mottled   *or*

mottled   dwarf   peach

- To confirm which of these two arrangements is correct, further mapping crosses need to be conducted involving these and other genes or chromosomal landmarks, such as a centromere.

Interference and the coefficient of coincidence

- A crossover decreases the likelihood of another crossover at a nearby site.
- This decrease is termed **interference** and results from a crossover in one region physically constraining the formation of another in the immediate vicinity.
- Interference has implications when analysing the results of trihybrid mapping crosses.
- Double crossover events are less common than expected, resulting in an underestimation of genetic distances.

- The **coefficient of coincidence** is a measure of crossover interference

$$\text{The coefficient of coincidence} = \frac{\text{number of observed double crossovers}}{\text{number of expected double crossovers}}$$

- In a trihybrid cross, the expected number of double crossovers is the product of the two single crossover frequencies.
- In the earlier tomato trihybrid mapping cross example, the expected proportion of double crossover progeny is $0.039 \times 0.12 = 0.00468$ (derived from 3.9% recombination between the *peach* and *mottled* genes, and 12.1% between the *peach* and *dwarf* genes).
- Expected number of double crossover progeny = expected proportion × observed number of progeny.
- In our example this equals $0.00468 \times 1002 = 4.68$ expected double crossover progeny. This compares with three observed double crossover progeny. Thus,

  the coefficient of coincidence $= \dfrac{3}{4.68} = 0.64$

- This means we observed 64% off the expected double crossovers.
- Interference = 1 − coefficient of coincidence.
- So the interference for our tomato three point cross is $1 - 0.64 = 0.36$
- This indicates that 36% of the expected double crossover progeny will not be produced.

Check your understanding

4.4 What is the genetic nature of a test cross and why is a test cross useful in genetic mapping?

4.5 Female fruit flies, heterozygous for cinnabar eyes (cn), plexus wings (px), and speck body (sp) were mated with male flies homozygous recessive for all three genes. Analysis of the progeny phenotypes indicated two genes were very close together. Which are the two genes and how close are they?

Progeny	px	sp	cn	3498	+	sp	cn	8
Phenotypes	px	sp	+	1410	+	sp	+	0
	px	+	+	1	+	+	cn	1489
	px	+	+	11	+	+	+	3483

4.5 MAPPING HUMAN GENES

Pedigrees are analysed for co-inheritance of a trait (often a disease) and alleles of a genetic marker. A logarithm of odds (LOD) score is calculated.

LOD scores

- A **LOD score** is a statistical test that assesses the likelihood of linkage between two genes.
- A LOD score compares the likelihood of obtaining the pedigree data under consideration if the two genes are linked to the likelihood of obtaining the same data by chance (i.e. as the result of independent assortment of genes on different chromosomes).
- Positive LOD scores indicate two genes are linked.
  - A LOD score of more than **3.0** is generally needed as evidence for linkage.
  - A score of +3 indicates 1000:1 odds that the association between the two genes observed in the pedigree(s) did not occur by chance, but because of linkage.
- Negative LOD scores indicate that linkage is unlikely.
  - A LOD score of less than −2.0 is considered evidence to exclude linkage.
- Scores of −2 to 3 are inconclusive. They indicate genes may be linked and more data need to be collected to confirm or refute linkage.
- Initially, producing LOD scores can seem complicated! However, only the principles of producing a LOD score need to be understood, as computer programs are routinely employed to analyse data.

Producing a LOD score from pedigree data

- Calculating a LOD score involves the following stages:
  - producing a pedigree
  - making a number of estimates of recombination frequency between the two genes under consideration
  - calculating a LOD score for each estimate
  - accepting the highest LOD score as the best indicator of linkage.
- Consider an example found in many textbooks: the co-inheritance of human ABO blood groups and nail patella syndrome—an autosomal dominant condition that results in poorly developed nails and kneecaps (Figure 4.9).

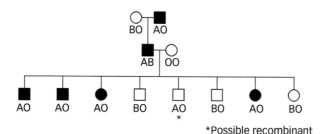

Figure 4.9 A three-generation pedigree showing inheritance of nail patella syndrome and ABO blood groups.

Mapping human genes

- Blood group allele **A** and nail patella allele **N** seem to be co-segregating in the pedigree of Figure 4.9 (consider individuals I-1, II-1, and III-1, 2, 3 and 7).
- Co-segregation of these two alleles is expected, but does not occur in individual III-5. This individual may be a recombinant.
- If the preceding two assumptions are correct, 1 of the 8 individuals of generation 3 are recombinants, giving a recombination frequency of 1/8 or 12.5% (and a map distance of 12.5 cM).
- A data set of eight individuals is small. Other explanations are possible:
  - the two genes could be a lot closer or more distant than the 12.5 cM suggested by this one pedigree
  - the two genes may not be linked, but on separate chromosomes, in which case, by chance, the chromosome with blood group allele A assorted into the same gamete as the chromosome with the nail patella allele seven out of a possible eight times.
- A LOD score is calculated to determine whether the blood type locus is linked to the nail-patella one (Table 4.5).
- The maximum LOD score was 1.099, when 0 = 0.125.
- This LOD score means that the possibility of linkage is about ten times more likely than the two loci assorting independently.
- A LOD score needs to have a value of ≥3 for confirmation that two genes are linked.
- When a LOD score is <3, data are gathered from other pedigrees and analysed.
- The highest LOD scores for each pedigree are summed.
- If this produces a final LOD score >3, this confirms linkage of the two genes.

(A)	LOD score $= \log \dfrac{[1-\theta]^{n-r}[\theta]^r}{[0.5]^n}$	θ = recombination frequency n = number of informative progeny r = number of recombinants
(B)	$\begin{aligned} \text{LOD score} &= \log_{10} \dfrac{[1-0.125]^{n-r}[0.125]^r}{[0.5]^n} \\[4pt] &= \log_{10} \dfrac{[0.875]^7[0.125]}{0.5^8} \\[4pt] &= \log_{10} \dfrac{0.0001917}{0.0000153} = 1.099 \end{aligned}$	

(C) Recombination frequency	0.05	0.1	0.125	0.15	0.2	0.25
LOD score	0.951	1.088	1.099	1.09	1.031	0.932

Table 4.5 Calculation of a logarithm of odds (LOD) score; (A) LOD score equation; (B) calculation of a LOD score for Figure 4.8 pedigree: recombination distance = 0.125, 12.5% recombination frequency; (C) LOD scores for a range of recombination frequencies

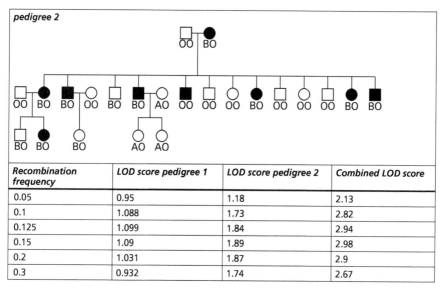

pedigree 2

Recombination frequency	LOD score pedigree 1	LOD score pedigree 2	Combined LOD score
0.05	0.95	1.18	2.13
0.1	1.088	1.73	2.82
0.125	1.099	1.84	2.94
0.15	1.09	1.89	2.98
0.2	1.031	1.87	2.9
0.3	0.932	1.74	2.67

Table 4.6 Logarithm of odds (LOD) scores from combined pedigree data. (A) Nail patella syndrome pedigree 2. All 18 individuals of generation II and III of pedigree 2 are informative; II-5, II-8, and III-1 are possible recombinants. Highest LOD scores from combined pedigree 1 and 2 scores are close to significance (LOD >3) in the range of 0.125–0.15. This indicates the loci for blood type and nail patella syndrome are likely to be 12.5–15 cM apart.

- The value of θ for which the LOD score is maximum is the best estimate of the distance between the two loci (Table 4.6).

Using DNA markers for human gene mapping

- It is rare to map a human gene relative to another functional gene.
- Instead, human genes are generally mapped relative to **DNA markers**.
- A DNA marker is a DNA segment with variable forms whose
  - chromosomal position is known
  - inheritance can be followed.
- Three types of DNA markers are used in gene mapping (Table 4.7).
- To be useful in mapping a DNA marker loci must:
  - have co-dominant alleles (so all alleles are detectable)
  - be highly polymorphic (so that segregation patterns are informative)—parents need to be heterozygous for the DNA marker in order to distinguish linkage phase in individuals.
- Microsatellites are often used to map human genes.
- Useful reference maps of the chromosomal location of thousands of microsatellites have been produced as the result of the human genome mapping project.

Mapping human genes

Marker	Description
Microsatellites or short tandem repeats	DNA segment composed of a short sequence (2, 3, or 4 nucleotides) repeated in tandem. Number of tandem repeats varies between individuals. See section 7.6 (p.134) for details of microsatellites.
Restriction fragment length polymorphisms	Variation in the DNA fragment sizes produced when DNA is cut with specific restriction endonucleases. See section 14.1 (p. 275) for details of restriction endonucleases.
Single nucleotide polymorphisms	Single nucleotide variation in DNA sequence. Two alleles are produced, e.g. individuals may have an A/T or C/G nucleotide pair at a specific site.

Table 4.7 DNA markers used in gene mapping

- These microsatellite maps are used in human gene mapping in a two-part process.
  - (i) An approximate chromosomal position is established by tracing the inheritance pattern of the gene to be mapped relative to a panel of ~200 microsatellites that cover the whole genome (~1 microsatellite per 7.5 cM). This low resolution mapping looks for co-segregation of a marker allele and the trait being mapped.
  - (ii) Detailed gene mapping with a second, smaller panel of microsatellites located close to the initial co-segregating microsatellite.
- To illustrate how microsatellites are used, consider again the nail patella locus mapped approximately 12.5 cM from the blood type gene on chromosome 9. To more precisely identify the chromosomal position of the nail patella gene:
  - ○ DNA from individuals of pedigrees 1 and 2 (Figures 4.5 and 4.6) is analysed for co-segregating alleles of a panel of microsatellites mapped to the nail patella region of chromosome 9
  - ○ for any microsatellites with alleles which are co-segregating with the nail patella syndrome LOD scores are calculated
  - ○ the data are often summarized as a graph (Figure 4.10)
  - ○ in Figure 4.9 microsatellites 1, 2, 3, and 4 are all linked to the nail patella gene, but at different distances
  - ○ microsatellite 3 is tightly linked to the gene determining nail patella syndrome.
- When a DNA marker is identified as tightly linked to a gene determining a trait the next step is to carry out physical mapping, i.e. the DNA nucleotide sequence in the region of the tightly-linked marker is examined for **candidate genes**.

Check your understanding

4.6 How are LOD scores calculated and how are they used in mapping human genes?

4.7 Interpret the following LOD scores and recombination frequencies (θ) for an autosomal dominant disease and allele 2 of a microsatellite marker.

θ	0.0	0.05	0.1	0.15	0.2	0.25	0.3	0.4	0.5
LOD	–	1.6	3.33	2.85	2.44	2.21	1.8	1.1	0.00

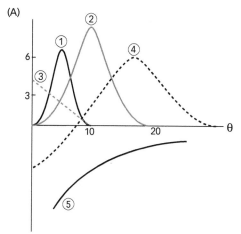

(A)

(B)

Microsatellite marker*	Maximum LOD score	Recombination frequency %	Map units: cM (between marker and gene)
1	5.8	2	2
2	7.9	4.5	4.5
3	3.8	0	0
4	6.4	12	12

*for the sake of simplicity, microsatellite identification numbers are shortened; e.g. 'microsatellite 3' = D9S315

Figure 4.10 Mapping the nail patella gene. (A) Logarithm of odds (LOD) score curves for five polymorphic microsatellite markers. Each curve represents the LOD score plotted against recombination frequency for a microsatellite with respect to the nail–patella locus. (B) Maximum LOD score data.

4.6 PHYSICAL MAPPING

- When a DNA marker is identified as tightly linked to a gene determining a trait the next step is to carry out physical mapping, i.e. the DNA nucleotide sequence in the region of the tightly linked marker is examined for candidate genes.
- This process is also termed **positional cloning**.
- Linkage analyse will have mapped the unknown gene for the trait of interest close to a DNA marker (generally less than 1 cM). Positional cloning proceeds as follows.
  - ○ Select relevant clones from a DNA library, which may be a genomic, chromosomal, or **cDNA** library.
  - (➔) *See section 14.1 (p. 279) for details of DNA libraries.*
  - ○ It may be necessary to order the cloned DNA segments (prepare a 'contig map') and sequence the cloned DNA segments.
  - (➔) *See section 14.4 (pp. 287–290) for details of DNA sequencing and contig mapping.*

Physical mapping

- ○ Analyse sequence for useful 'landmarks', e.g. promoter consensus sequences, start and stop codons in the same reading frame, and intron splice sites.
- It may be appropriate to perform a **zoo blot**. If the DNA sequence is known for a gene in a related species, this can be used to prepare a probe which will hybridize to, and thus identify, the relevant gene in the species under analysis.
- Once one or more candidate genes have been identified, the sequences are compared in individuals of different phenotypes. For example, are specific, mutant sequences only found in individuals expressing the disease under investigation?

 Check your understanding

4.8 You recently isolated a mutant in the fruit fly, *Drosophila melanogaster*, that results in the flies being unable to orientate themselves in a straight line, i.e. they continuously fly in circles. You have called this new mutation **cir**. You suspect that the **cir** mutation is located on chromosome 3 close (within 2 cM) of the striped body gene **str**. Discuss fully the genetic crosses and subsequent analyses you would do to investigate your hypothesis.

online resource centre You'll find guidance on answering the questions posed in this chapter—plus additional multiple-choice questions—in the Online Resource Centre accompanying this revision guide. Go to **www. oxfordtextbooks.co.uk/orc/thrive/** or scan this image:

5 Quantitative Genetics

Quantitative genetics analyses the inheritance of complex traits.

5.1 CONTINUOUS TRAITS

- In genetics we make the distinction between:
  - **continuous traits**, analysed by the tools of quantitative genetics (described in this chapter)
  - **discontinuous traits**, analysed by the tools of Mendelian genetics (described in Chapters 2 and 3).
- Key features of continuous and discontinuous traits are compared in Table 5.1.
- A continuous trait displays a large number of possible phenotypes, which can be difficult to distinguish, e.g. human height.
- Generally, there are many intermediate forms between two extreme phenotypes.
- Examples of continuous traits are:
  - weight, height, and blood pressure in humans
  - growth rate and milk production in cattle
  - seed weight and height in plants.
- Continuous traits are **multifactorial**, i.e. they are determined by both:
  - segregating alleles of many different genes
  - various environmental factors.
- The relationship between genotype and phenotype is complex. For example:
  - alleles of different genes often interact with each other
  - varying environmental factors can result in a single genotype producing a range of phenotypes.
- Many continuous traits of animals and plants are important agriculturally and in animal farming.
- Thus, it can be important to assess how much of the phenotypic variation of continuous traits is attributable to genetic influences, as this determines the outcome of selective breeding programmes.

	Discontinuous traits	*Continuous traits*
	Qualitative expression: phenotypes fall into a few discrete classes, e.g. normal or sickle haemoglobin	**Quantitative** expression: phenotypes show a wide range, e.g. human blood pressure
Genetic control	One or a few genes	Multiple genes (or **polygenic**)
Environmental influence	Absent/minimal	Strong
Relationship between genotype and phenotype	Predictable: one phenotype/ genotype	Unpredictable: different phenotypes produced by a given genotype
Mode of analysis	Either/or (expressed/not expressed)	Trait measured, weighed, or counted
	Phenotypic proportions are scored among offspring of controlled genetics crosses between selected individuals	Statistical analysis of phenotypes among individuals of a randomly sampled population

Table 5.1 A comparison of discontinuous (Mendelian) and continuous (quantitative) traits

5.2 THE ROLE OF ADDITIVE GENES

- The expression of many genes determines the phenotype of a continuous trait.
- A gene's effect may be **additive**, dominant, or epistatic.
- Generally, additive interactions are the most important in expression of a continuous trait.

  ➜ *See Chapter 3 (pp. 38 and 47) for discussions of dominance and epistasis.*

Additive genes and possible phenotypes

- Each allele of an additive gene has a quantifiable contribution to the final phenotype.
- Certain alleles increase, while others decrease, expression of a quantitative trait; e.g. height, colour intensity, and size.
- The expression of a trait is the sum of all additive alleles.
- Alleles of additive genes are incompletely dominant.
- Predictions of the number of different possible genotypes can be made from considering Mendelian genetics:
  ○ three different genotypes result when a trait is determined by one gene with two incompletely dominant alleles (AA/Aa/aa)
  ○ thus, 3^n (where n=number of genes) predicts the number of possible genotypes when each gene has two incompletely dominant alleles (Table 5.2)
  ○ the number of resulting phenotypes is likely to be greater through interactions of environmental factors with each genotype.
- Consider a situation where two genes (*A* and *B*), each with two incompletely dominant alleles, contribute to the length of an insect species' wing (Table 5.3).

Number of genes	1	2	3	5	10
Number of genotypes	3	9	27	243	59,049

Table 5.2 Predicting genotype numbers (each gene possesses two alleles)

Genotype	*Insect wing length*	*F2 numbers*
AABB	12	1
AABb	11	2
AAbb	10	1
AaBB	10	2
AaBb	9	4
Aabb	8	2
aaBB	8	1
aaBb	7	2
aabb	6	1

Table 5.3 Illustrating the effect of additive genes: determining insect wing length. (In this hypothetical example, alleles **A** and **a** contribute 4 mm and 2 mm, and alleles **B** and **b** contribute 2 mm and 1 mm to wing length.)

Statistical analysis of continuous traits

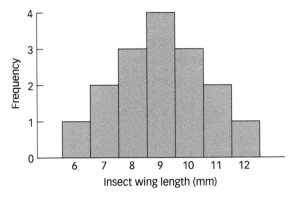

Figure 5.1 Histogram showing the distribution of insect wing length in the F2 generation of a Mendelian cross (parents (AABB and aabb; F1 AaBb)).

- The histogram of the insect wing length data in Figure 5.1 shows seven categories (phenotypes); it has the bell-shaped curve indicative of a normal distribution.
- With more additive genes involved in a trait many more phenotype categories result. Environmental influences can modify expression of each genotype and so further add to the number of measurable categories, which are best represented by a continuous curve (Figure 5.2).
- The study of continuous traits requires the analysis of a large number of individuals. Thus, the expression of most quantitative traits is best described and analysed statistically.

> ### ⇒ *Check your understanding*
>
> 5.1 How do continuous and discontinuous traits differ?
> 5.2 Explain why there is often a complex relationship between genotype and phenotype for continuous traits.
> 5.3 Why do continuous traits often show many phenotypes?

5.3 STATISTICAL ANALYSIS OF CONTINUOUS TRAITS

- As quantitative traits are measured on a continuous scale, **statistical methods** give the best ways of describing and analysing their patterns of inheritance.
- Statistical methods are also appropriate tools of analysis as quantitative genetics considers the phenotypic expression of large groups of individuals (**populations**) through investigating a subgroup or **sample.**

The use of the normal distribution

- The range of phenotypes shown by a continuous trait are presented as a **frequency distribution** (Figure 5.1), i.e. a histogram of the numbers of individuals (y-axis) showing the different phenotypes (x-axis).
- Typically, quantitative traits exhibit a symmetrical **normal distribution**. The bell-shaped curve of a normal distribution is produced by connecting the points of a frequency distribution with a line (Figure 5.2).
- A normal distribution is common when a large number of independent factors contribute to a measurement. This is the case with continuous traits that are influenced by multiple genes and environmental factors.
- A normal distribution is described by various parameters:
  - mean
  - variance
  - standard deviation.
- The **mean** locates the centre of the distribution of phenotypes.
- It is calculated by summing all the individual measurements and dividing by the total number of measurements in the sample (Table 5.4).
- The **variance** and **standard deviation** give useful information about the variability of a trait, i.e. whether there is a narrow range of phenotypic expressions clustered around the mean, or a wide range of phenotypes.
- The variance is a measure of how far each value in a data set deviates from the mean (Table 5.5).

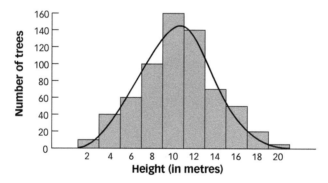

Figure 5.2 Histogram showing variation in tree height in Mikumi National Park, Tanzania (576 trees were measured; the height range was divided into 10 classes).

Formula	Terms
$\bar{x} = \dfrac{\Sigma x_i}{n}$	$\bar{x}$=mean Σ=sum of x_i=each sample n=sample number

Table 5.4 Calculating a mean value

Statistical analysis of continuous traits

Formula	Method of calculation
$s^2 = \dfrac{\Sigma(x_i - \bar{x})^2}{n-1}$	• The mean is subtracted from each measurement and the resulting value squared • These squared deviations are summed • This summed total is divided by the number of original measurements minus 1

Table 5.5 Calculating the variance

- The usefulness of knowing the variance (s^2) is illustrated in Figure 5.3.
- The standard deviation (SD) is another useful parameter for characterizing a normal distribution.
- In a normal distribution various percentages of the sample fall within certain ranges determined by the SD:
  - 66% of the measurements of a normal distribution lie within +1 or −1 SD
  - 95% within 2 SD
  - 99% within 3 SD.
- The SD = $\sqrt{s^2}$ (i.e. the square root of the variance).

Correlation and regression

- The mean, variance, and SD describe an individual trait.
- Quantitative geneticists often want to know if there is a relationship between a pair of traits.
- They look for **correlations** between traits, i.e. whether a change in one trait is associated with a particular trend in another. For example, a large avian study in the 1960s showed that egg weight increased as female body weight increased.
- A **correlation coefficient** (r) is calculated to assess a possible relationship between two traits.

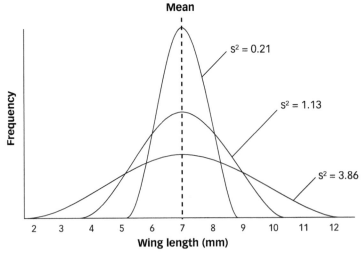

Figure 5.3 The usefulness of the variance. The three data sets have the same mean, but different variances.

- It is calculated using the statistics that describe the variability of a trait, i.e. the variance and SD:
  - firstly, a **covariance (cov)** is determined:

$$cov = \frac{\Sigma(x_i - \bar{x})(y_i - \bar{y})}{n-1}$$

  where x_i are the individual measures and $\bar{x}$ the mean of one trait; y_i and $\bar{y}$ are the readings and mean for the second trait; and n=sample size
  - secondly, the covariance is divided by the product of the SD for each trait:

$$\text{Correlation coefficient } (r) = \frac{cov_{xy}}{S_x S_y}$$

- The value of the correlation coefficient (often called the 'r-value') can vary between -1 and $+1$.
- It gives two pieces of information.
  - Its sign (positive or negative) indicates the nature of the correlation:
    - a positive r-value indicates a direct correlation—as one variable increases so does the other
    - a negative r-value indicates an inverse relationship—as one variable increases the other decreases
  - Its numerical value measures the strength of association between the two variable phenotypes:
    - the closer to $+1$ or -1, the stronger the correlation (Figure 5.4)
    - coefficients near 0 indicate weak or no correlation.
- It is important to remember that a correlation between two parameters does *not* confirm a causal relationship, but certainly indicates that further investigation is worthwhile.
- If an association has been established between two phenotypes, it is useful to be able to predict the precise expression of the second phenotype if one has information about the first.
- This prediction is achieved by plotting a **regression line** (calculated best fit-curve). In Figure 5.4 lines are best-fit by eye.
- A regression line is established mathematically:

$$y = a + bx$$

  - **a** represents the y intercept of the regression line, i.e. the value of y when x=0
  - **b** represents the regression coefficient calculated from dividing the covariance of x and y, by the variance of x:

$$b = \frac{cov_{xy}}{S_x^2}$$

Statistical analysis of continuous traits

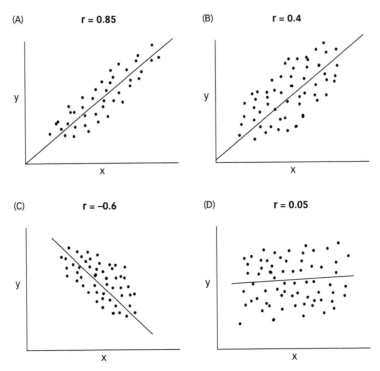

Figure 5.4 Scatter plots showing different types of correlation (x=independent variable, y=dependent variable, r=correlation coefficient). (A) and (B) show a positive correlation, indicating a direct association between the two variables. (C) Shows a negative correlation and (D) shows that there is no association between the two variables.

- o a can be calculated by substituting the regression coefficient and mean values of x and y into the following:

$$a = \bar{y} - b\bar{x}$$

- Once values for **a** and **b** are calculated, the line is fully established by substituting values of x in the regression equation, y=a+bx. An example is given in Figure 5.5.
- As well as analysing a possible association between two variables in a single group of individuals, the correlation and regression coefficients can be used to assess a single variable in different groups of individuals.
- For example, these coefficients are commonly used to assess whether there is a significant correlation of expression of a particular phenotype in parents and their offspring.

Check your understanding

5.4 What information does (i) the mean and variance, and (ii) the correlation coefficient provide about a sample?

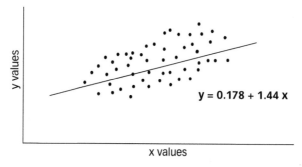

Figure 5.5 Plotting a regression line.

5.4 HERITABILITY

- **Heritability** is the proportion of phenotypic variation in a population that arises as a result of genetic differences between individuals.
- Knowing the heritability of a trait among a group of individuals has important practical applications. For example, it informs breeding programmes in plants and animals, or the way we approach treating certain human diseases.
- The variance of a trait can be used to gain an indication of the relative input of genetic and non-genetic factors into determining a character.

Phenotypic variance

- From the variable expression of a character a mean and variance can be calculated (see section 5.3). This variance is the **phenotypic variance**, which is represented by V_P.
- Some of this phenotypic variance (variation in phenotypic expression) will result from differences in genotypes among the individuals of the study population. These differences contribute the **genetic variance** (V_G).
- Variable environmental factors account for other differences between individuals and contribute the **environmental variance** (V_E).
- Phenotypic variance can therefore be apportioned into two main components:

$$V_P = V_G + V_E$$

- We sometimes recognize a third component to phenotypic variance: a genetic–environmental variance, resulting from interaction between genetic and environmental factors:

$$V_P = V_G + V_E + V_{GE}$$

Heritability

- Genetic variance can be further subdivided:

$$V_G = V_A + V_D + V_I$$
$$V_A = \text{additive genetic variance}$$
$$V_D = \text{dominance genetic variance}$$
$$V_I = \text{epistatic genetic variance}$$

- Therefore:

$$V_P = (V_A + V_D + V_I) + V_E$$

Types of heritability

- We calculate either:
  - **broad sense heritability**, which gives a general assessment of genetic versus environmental input to phenotypic expression of a trait
  - **narrow sense heritability**, which focuses on the input of additive genes.

(i) Broad sense heritability

- Broad sense heritability (H^2) represents the proportion of phenotypic variance that is due to total genetic variance:

$$H^2 = \frac{V_G}{V_P}$$

- Values for H^2 can range from 0 to 1.
- A value of 1 indicates all of the phenotypic variance results from genotype differences. Conversely, a value of 0 indicates that only environmental factors are influencing phenotype.
- Typically, both genetic and environmental factors influence expression of a trait, resulting in a heritability value between 0 and 1.

(ii) Narrow sense heritability

- Narrow sense heritability (h^2) represents the proportion of phenotypic variance that is due to additive genetic variance:

$$h^2 = \frac{V_A}{V_P}$$

- Once again, values range from 0 to 1, the higher the value indicates greater genetic input.
- Narrow sense heritability is particularly useful to breeders, as it indicates the degree to which offspring resemble parents (Section 5.5).

Calculating heritability

- There are various ways of calculating heritability:
  - (i)   Measuring phenotypic variance when one component is eliminated
  - (ii) Comparing the resemblance of parents and offspring
  - (iii) Comparing the V_P of individuals with varying degree of relatedness
  - (iv) Measuring the response of individuals to selection.

It is important to remember that heritability calculations are for a specific group of individuals in a specific environment. Values will change for different populations and among environments.

(i) Broad sense heritability by elimination of variance components

- To calculate H^2 we need values for V_P and V_G:

$$H^2 = \frac{V_G}{V_P}$$

- A value for V_P is gained easily, as it represents the overall variability of phenotypic expression.
- A value is determined for V_G by running two sets of experiments.
  - (i)   a trait is measured in a population of genetically varied individuals under a defined set of environmental conditions. The range of phenotypic expression produces a value for V_P.
  - (ii) A population of genetically identical (homozygous) individuals is produced by cloning or generations of inbreeding. As $V_P = V_G + V_E$ and we have eliminated V_G, as all individuals share the same genotypes, V_P in this population $= V_E$.
- We now have values for V_P and V_E, so:

$$V_G = V_P - V_E$$

- An example of calculating broad sense heritability is given in Box 5.1.
- The method of calculating heritability, described in Box 5.1, relies on being able to create genetically identical organisms, which is time consuming (and clearly impossible in humans).
- This method also assumes V_E is the same for genetically identical and variable individuals. But different genotypes are likely to respond differently to the same environment. Hence, there is also a V_{GE}, although this is generally ignored in these calculations.

(ii) Narrow sense heritability by regression analysis

- If there is an inherited component to the expression of a trait, offspring should resemble their parents more than they do unrelated individuals.

Heritability

Box 5.1 Calculating broad sense heritability

The American plant geneticist Edward East (1879–1938) studied the inheritance of corolla tube length in the tobacco plant *Nicotiana longiflora*.

Corolla tube

- East established pure-breeding short- and long-tubed varieties.
- He measured the variation in corolla tube length in these two 'parental' groups and in the F1 generation when these two varieties were crossed.
- Each parental strain and the F1 should be genetically identical.
- Parental strain A is homozygous for short tube genes, strain B homozygous for long tube genes, and F1 individuals are heterozygous.
- Thus, any variation among parental and F1 individuals will be the result of environmental factors.
- This gives three estimates for V_E. East found an average value of 8.76.
- The variation in corolla tube length was much greater in the F2 generation. The variance equalled 40.76. F2 individuals differ in genotypes, as well as experienced variable environmental factors. So this variance$=V_P$.
- These results provide sufficient data to calculate broad sense heritability (H^2).
- $V_G=V_P-V_E$, so $V_G=40.76-8.76=32.2$.
- $H^2=32.2/40.76=0.79$.
- This heritability value suggests a strong genetic input to the expression of corolla tube length in tobacco.

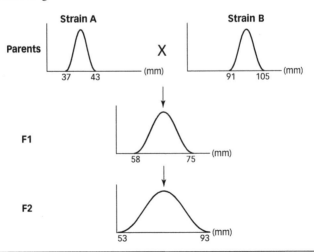

- The strength of any association is assessed by plotting a regression line (Figure 5.3).
- The regression coefficient (**b**) equals the narrow sense heritability, as the additive genetic variance is responsible for resemblance between parents and offspring:

$$h^2 = b$$

- This method involves:
  - Measuring a character in many families
  - Plotting the mean parental phenotype against the mean offspring phenotype for each family
  - Thus, each data point represents one family with the mean parental phenotype on the x-axis and mean offspring measure on the y-axis (Figure 5.6)
  - The value of the regression coefficient (b) lies between 0 and 1; the higher its value, the greater the magnitude of the heritability of the trait under consideration.
- If the phenotype of only one parent is known, the narrow sense heritability = twice the regression coefficient ($h^2 = 2b$).

(iii) Broad sense heritability from degrees of relatedness
- This method is described in section 5.5, page 97.

(iv) Narrow sense heritability and response to selection
- Individuals of a species become genetically suited to their environment through natural selection.
- Humans have practised **artificial selection** for many centuries. Agricultural plants and animals with desired traits are selected for breeding.
- The amount by which a quantitative trait changes in a single generation of selective breeding, i.e. its **response to selection (R)**, is determined by two key factors:
  - **selection differential (S)**, i.e. the difference between the average phenotype of the selected individuals (for breeding) and the average population phenotype
  - narrow sense heritability (h^2) of the trait.

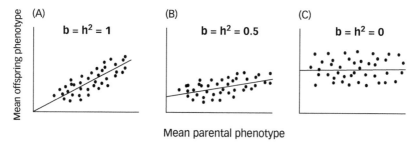

Figure 5.6 Using the regression coefficient (b) to determine heritability (h). (A) An inherited trait, solely determined by genetic factors. (B) Both genetic and environmental factors determine phenotype—in this example equally. (C) There is no relation between parental and offspring phenotypes; only environmental factors influence the trait's expression.

Heritability

- The relationship between these three factors is defined as follows:

$$R = h^2 S$$

- To illustrate a trait's response to selection, consider a poultry farmer maximizing his profits. He wants hens that lay for the maximum possible days each year. Thus, he breeds from those hens that laid the greatest number of eggs the previous year.
- The response to selection of this trait will depend upon:
  - the narrow sense heritability of egg-laying in the farmer's population of hens, i.e. the extent of the genetic input to egg-laying
  - how stringent the breeder is in selecting his individuals for breeding (the selection differential), e.g. does the poultry farmer select the top 2% or 10% highest egg-laying hens for breeding?
- Figure 5.7 shows the calculation of a response to selective breeding.
- The fact that there has been a response to selection, i.e. an increase in egg yield as a result of selective breeding, indicates that there is a genetic input to this trait.

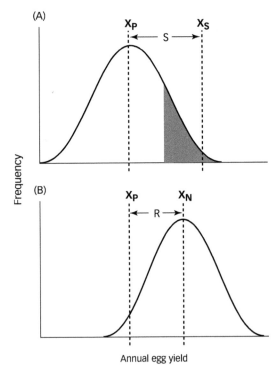

Figure 5.7 Calculating the response to selection. (A) Distribution of annual egg yield in the parental population. Hens that fall in the shaded part of the graph are selected for the breeding programme. (B) Egg yield from offspring of selected hens.

- An estimate of this genetic input, i.e. the heritability (h^2), comes from considering the selection response relative to the selection differential:

$$h^2 = \frac{R}{S}$$

- Note that it is the narrow sense heritability that is calculated.
- Once a value for the heritability of a trait is known for a population, then the response to a selective breeding programme can be predicted for a given selection differential ($R = h^2 \times S$).
- There will be limits to a selective breeding programme. After many generations of selective breeding, a trait will cease to respond to selection because all individuals will have become homozygous for alleles encoding the selected trait.
- It is common to find **correlated responses** to selection. When one trait is selected, others change at the same time. The associated change might not be advantageous.
- For example, selection for fast growth is a common goal of selection. In cattle this might also result in larger calves being born, which can cause difficulties during calving.

Important points about heritability

- It is important to be clear on the kinds of information that heritability values do and do *not* give.
- A heritability value is specific to the population you are analysing.
- It applies to a population, not an individual. For example, the heritability for milk yield in a large herd of cattle on a farm is 0.57. This indicates that 57% of the variability in milk yield among this herd is the result of genetic factors. It does not mean that for any cow from this herd (or any other) genes determine 57% of the milk yield.
- Heritability does *not* define the degree to which a character is genetically determined. It measures the proportion of the phenotypic variation that is the result of genetic differences between individuals for a particular population of individuals under a given set of environmental conditions.
- A heritability value needs to be calculated afresh for each population and each environment.
- To emphasize these points consider a much quoted example regarding human height.
  - Nutrition and health are important factors influencing height:
    - in wealthy, developed countries, where most people have a good diet and health care, heritability for height is high
    - in developing countries, where there is often a higher proportion of malnourished individuals who suffer from diseases affecting height, a measure of heritability is lower

- genetic factors might be the same in both populations, but there is a greater input of environmental factors in the second, thus heritability values may be very different.
- However, on the positive side, there does tend to be general agreement on heritability values for a given trait in different populations!
- Indeed, in the early twenty-first century, with the development of new molecular biology tools to analyse DNA, decades of heritability studies have indicated a significant genetic input to many quantitative traits and justify searching for the relevant genes.

  ➤ *See section 5.6 (p. 99) for details of mapping quantitative trait loci.*

Looking for extra marks?

The Illinois long-term selection experiment for oil and protein in maize is the world's longest running plant genetic experiment. It started in 1896 when four strains were established: Illinois high and low oil, and high and low protein. Selection is in the direction of the strain's name. It has created 12 populations that vary significantly in oil and protein content, and are continuing to respond to selection.

 Check your understanding

5.5 List and define the different components of the phenotypic variance (V_P).

5.6 What is the difference between narrow sense and broad sense heritabilities?

5.7 List different ways by which heritabilities can be calculated

5.8 Why does a population's response to selection generally decline after many generations of selection?

5.5 HUMAN QUANTITATIVE TRAITS

- Geneticists recognize two types of human quantitative trait:
  - continuously expressed traits, such as height or skin colour
  - discontinuously expressed or **threshold traits**, such as cancer and schizophrenia; the quantitative aspect is liability to express such traits.

Threshold traits

- Threshold traits are inherited quantitatively, but expressed qualitatively, i.e. the trait is either present or absent.
- Threshold traits include many adult-onset diseases (e.g. diabetes mellitus, cancer, epilepsy, hypertension, manic depression, schizophrenia, Alzheimer's disease) and

congenital malformations (e.g. cleft palate, congenital dislocation of the hip, congenital heart disease, neural tube defects, pyloric stenosis, talipes).

- The quantitative aspect to threshold traits is an individual's susceptibility to expression (also referred to as liability or risk).
- Susceptibility is multifactorial; it is determined by multiple genes and various environmental factors.
- Certain alleles of various genes and particular environmental factors predispose expression. If sufficient of these susceptibility factors are present for an individual, he/she develops the disease or condition.
- The predisposing genetic and environmental factors can be assigned values which can be scored in a population. Individuals' scores show a normal distribution (Figure 5.8A).
- Expression of a threshold trait occurs when a certain threshold susceptibility is exceeded.
- There are some conditions where males and females show different threshold values for expression (Figure 5.8B).
- Examples of conditions with:
  ○ lower male threshold include pyloric stenosis, infantile autism, and multiple sclerosis
  ○ lower female threshold include congenital hip dislocation.

Check your understanding

5.9 Consider a threshold trait which is twice as common in females as males. What type of parents, and why, are at higher risk of producing an affected child: affected father/unaffected mother or affected mother/unaffected father?

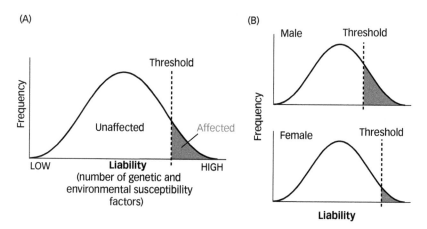

Figure 5.8 (A) Expression liability of a threshold trait. (B) Different male and female liability threshold values (here for pyloric stenosis).

Human quantitative traits

Measuring heritability

- For obvious reasons, selective breeding experiments to calculate the response to breeding or the variance components (as described in the previous section) cannot be conducted with humans to determine heritability.
- Instead, twin studies are used to assess the relative input of genetic and environmental factors.
- Most pairs of twins share the same general environment, but there is a difference in genetic similarity:
  - **monozygous**, MZ (or identical) twins are genetically identical, as they originate from a single zygote
  - **dizygous**, DZ (or non-identical) twins share 50% of their genes as they are the result of two separate fertilization events.
- To assess the genetic input to expression of human quantitative traits comparisons are made between their expression in monozygous and dizygous twins.
- Heritability values are calculated from:
  - correlation coefficients for continuous quantitative traits
  - **concordance rates** for threshold traits.

(i) Concordance rates

- Twins are termed:
  - **concordant** for a trait if both or neither of a pair express it
  - **discordant** if only one twin shows the trait.
- Concordance values (i.e. how often trait expression is concordant) are calculated for MZ and DZ twins.
- If there is a genetic input to expression of a trait, concordance rates will always be higher in monozygous than dizygous twins.
- For a trait totally determined by genotype, concordance rates should be 1.0 for MZ twins and 0.5 for DZ twins.
- More common are intermediate figures indicating input of both genetic and environmental factors (e.g. schizophrenia, Table 5.6)
- Similar high concordance rates in both MZ and DZ twins suggest a strong environmental input to a trait's expression, e.g. expression of an infectious disease (measles, Table 5.6).
- A comparison of concordance rates in MZ and DZ twins can be used to give a measure of the heritability of a trait (i.e. the proportion of the phenotypic variance caused by genetic factors).
- Heritability is calculated from the following formula:

$$h = 2(C_{MZ} - C_{DZ})$$

where C_{MZ} = the concordance rate for MZ twins and C_{DZ} = the concordance rate for DZ twins.

Trait	MZ	DZ	Heritability
Alcoholism	0.6	0.3	0.6
Blood pressure (systolic)	0.55	0.25	0.6
Body mass index	0.95	0.53	0.84
Cleft lip/palate	0.38	0.08	0.6
Club foot	0.32	0.03	0.58
Diabetes mellitus (type 1)	0.5	0.1	0.7
Diabetes mellitus (type 2)	0.9	0.2	0.85
Height	0.94	0.44	1.0
Measles	0.95	0.87	0.16
Multiple sclerosis	0.28	0.03	0.5
Schizophrenia	0.47	0.12	0.7
Spina bifida	0.72	0.33	0.78

Table 5.6 Concordance rates or correlation coefficients and resulting heritability values (summary of many studies) in monozygous (MZ) and dizygous (DZ) twins for selected human continuous and threshold traits

(ii) Correlation coefficients

- For a continuous quantitative trait, measurements of that trait are made in each twin, and a **correlation coefficient** is calculated for both MZ and DZ samples.
  - ➡ *See section 5.3 (p. 85) for calculation of a correlation coefficient.*
- For a trait totally determined by genes, the value of the correlation coefficient for monozygous twins would be double its value for dizygous twins; a value of 1.0 compared to 0.5.
- A correlation coefficient of 1.0 indicates each pair of twins expresses the trait in exactly the same way, e.g. have the same height.
- Heritability is calculated from the same formula as used with concordance values:

$$h = 2(r_{MZ} - r_{DZ})$$

  where r_{MZ} = the correlation coefficient for MZ twins and r_{DZ} = the correlation coefficient for DZ twins.
- Concordance rates and correlation coefficients produced by analysing a trait in other relatives can also be used to calculate heritability values, e.g. expression of a trait in parents and their children.

Limitations of twin studies for heritability estimates

- An assumption is made that the environments of MZ and of DZ twins are equally similar. There are, however, various factors that result in environmental experience being more similar for a pair of MZ than DZ twins.
- MZ twins:
  ○ are often treated more similarly than DZ twins
  ○ develop exceptionally close bonds with each other.

- The uterine environment differs for MZ and DZ twins. Each DZ twin has its own amnion and chorion. MZ may share both, one, or neither.
- The generally greater similarity in the environment of MZ compared with DZ twins results in higher concordance rates for MZ twins, inflating the apparent influence of genes on a trait.
- One way to eliminate this MZ environmental bias is to study twins that are raised apart, in separate environments. Concordance should be solely due to genetic factors.
- The limitations of such studies are:
  - small sample sizes
  - some contact between twin pairs prior to or during study period.

> ### Looking for extra marks?
> The Minnesota Twin Study is a long-term research project assessing genetic and environmental influences on development and psychological traits. Participants include MZ and DZ twins, and adoptive and biological siblings and their parents.

Adoption studies

- Given the small number of twins who are raised separately, broader adoption studies have been conducted over the years to assess genetic contribution to quantitative traits, particularly disease.
- Expression of a trait is compared in the following two groups:
  - adopted children and their biological parents
  - adopted children and their adoptive parents.
- In many cases adoptive children have been found to express diseases more often than children in comparative control groups.
- For example, the rate of schizophrenia among adopted children with a biological parent who has schizophrenia is found repeatedly to be 8–10%, compared to 1% of adopted children of non-affected biological parents, suggesting a genetic input to the development of schizophrenia.
- However, as with twin studies, there are factors that can inflate an apparent genetic influence:
  - prenatal environmental influences
  - children living with biological parents during early childhood years
  - matching adoptive and natural parents in terms of socio-economic and other factors.

5.6 QUANTITATIVE TRAIT LOCI

- **Quantitative trait loci** (QTLs) refers to genes or chromosomal regions (with as yet unidentified genes) that contribute to the expression of quantitative traits.

- There are many factors that make it difficult to identify QTLs, such as:
  - locus heterogeneity—the fact that many genes determine quantitative traits, with each gene having a small effect
  - interactions between genes and environmental factors
  - genetic issues, e.g. variable penetrance and age-dependent expression.
- Twenty-first century developments in molecular analysis of DNA (e.g. the use of microarrays) have significantly progressed the identification of QTLs.
- There are three main QTL mapping methods:
  - **linkage analysis**
  - affected sibling–pair analysis
  - **association studies**.
- The following sections describe mapping of human QTLs involved in expression of threshold traits. When mapping QTLs involved in continuous traits it is best to analyse individuals with extreme values of expression.

Linkage analysis

- This is the traditional gene mapping approach.
  - ➜ *See section 4.5 (p. 75) for details of mapping the locus of a single human genetic condition.*
- Disease-expressing families are identified.
- Linkage analysis is undertaken with a large panel of polymorphic DNA markers
- If a logarithm of odds score of ≥3 is obtained with one of these markers, it is assumed that the chromosomal region around the marker contains a gene of interest.
- Positional cloning identifies a candidate gene.
- This approach has successfully identified human QTL, e.g. the human breast cancer susceptibility gene *BRCA1*.
- However, linkage analysis tends to identify genes with a relatively strong contribution to a quantitative trait.
- The following two methods have identified many more human QTLs.

Affected sibling pairs analysis

- A large number of sibling pairs that both express a quantitative condition (**affected sibling pairs**) are scanned for shared alleles of polymorphic markers.
- If there is a significantly increased sharing of the same allele of a marker above the expected 50% for siblings, this indicates the DNA marker is close to a QTL.
- Positional cloning can then identify a candidate gene.
- This approach identified that human HLA alleles D3 and D4 increased susceptibility to type I diabetes.

Quantitative trait loci

Genome-wide association studies

- This approach is being used increasingly to identify human QTLs, made possible by the development of high throughput molecular techniques that can accommodate the necessary huge sample sizes.
- Thousands of polymorphic DNA markers are scanned in thousands of expressing and matched non-expressing individuals.
- For example, microarrays can assay one million single nucleotide polymorphisms (SNPs) in a collection of case/control individuals.
- As with the other two mapping approaches, the scan looks for an association of DNA marker alleles with expression of the quantitative trait.
- Again, positional cloning identifies candidate genes and expressing alleles.
- A given allele may well be present in non-expressing individuals; critically, such an allele occurs at a higher frequency among those individuals expressing a quantitative trait.
- Table 5.7 shows the HLA-B27 allele is associated positively with expression of ankylosing spondylitis.
- The human **Hapmap Project** is helping to increase the efficiency of association studies.
- This project identifies SNPs linked together on a chromosome into a **haplotype**.
- This enables fewer SNPs to be assayed and genotyped in an association study: alleles identified at one **tag SNP** predict alleles present in a group of SNPs.
- Thus, ~500,000 tag SNPs are genotyped instead of 10 million.
- In the early 2000s, this approach identified four new breast cancer susceptibility genes.

	Expressing AS	*No AS*
HLA-B27 present	95	700
HLA-B27 absent	5	92,900

Table 5.7 Study of association between expression of ankylosing spondylitis (AS)* and HLA-B27

*Ankolysing spondylitis is a form of arthritis which primarily affects the sacroiliac joints. Inflammation leads to ossification and fusion of joints. In this hypothetical study, allele HLA-B27 is × 12.67 (0.95/0.075) more frequent in AS individuals.

 Check your understanding

5.10 How are genes that influence human quantitative traits identified?

online resource centre You'll find guidance on answering the questions posed in this chapter—plus additional multiple-choice questions—in the Online Resource Centre accompanying this revision guide. Go to www.oxfordtextbooks.co.uk/orc/thrive/ or scan this image:

6 The Genetics of Bacteria, Viruses, and Organelles

Research using bacteria and viruses initiated the field of molecular genetics. They remain essential tools for probing the nature of genes and gene expression.

Key concepts

- Bacteria and viruses have small haploid genomes.
- They are well suited to genetic studies because they have high rates of reproduction and produce large numbers of progeny.
- Plasmids are often present in bacterial cytoplasm.
- DNA can be transferred between bacterial cells by conjugation, transformation, or transduction.
- Viruses have DNA or RNA genomes.
- Bacteriophages are DNA viruses that infect bacteria.
- Retroviruses are RNA viruses that infect eukaryotic cells.
- Mitochondria and chloroplasts have their own genetic systems.

6.1 THE BACTERIAL GENOME

- Most bacteria have a single circular chromosome, several million nucleotides in length, e.g. *Escherichia coli* (strain K12) consists of 4.64 million nucleotides, encoding 4377 genes.

The bacterial genome

- A few bacteria contain multiple chromosomes. For example:
  - *Vibrio cholerae* (causing the disease cholera) has two circular chromosomes; one of these chromosomes contains genes encoding essential functions, while the other contains genes involved in virulence
  - the nitrogen-fixing bacterium *Rhizobium melilotii* has three circular chromosomes.
- Linear chromosomes are found in a few bacterial species, e.g. *Borrelia burgdorferi* (causing Lyme disease).
- The bacterial chromosome(s) is packaged within a distinct region of the cytoplasm termed the **nucleoid body**.

Packaging of bacterial DNA

- The length of a bacterial chromosome is several orders of magnitude greater than a bacterial cell, so it needs to be packaged.
- Bacterial DNA is condensed into the nucleoid region by **supercoiling**.
- Supercoils are produced when a strain (in the form of extra helical turns) is placed on the DNA helix. This strain causes the helix to twist (Figure 6.1).
- **Positive supercoils** are produced when a DNA helix is over-rotated and **negative supercoils** when a helix is under-rotated.
- To understand how supercoiling compresses DNA into a smaller volume:
  - gently twist a rubber band so that it forms a few coils
  - now twist it further so that the original coils fold over one another to form a condensed ball.
- Enzymes called **topoisomerases** create the supercoils. They add or remove rotations from the DNA helix by transiently breaking the nucleotide strands, rotating the ends around each other, and then resealing the ends.

Prokaryotic genomes	Eukaryotic genomes
Most prokaryotes contain a single circular chromosome	Eukaryotes contain multiple linear chromosomes
Located in the nucleoid region of the cytoplasm	Located in a membrane-bound nucleus
Supercoiling compacts the DNA	DNA is condensed via histones into nucleosomes and chromatin
One copy per cell: prokaryotes are haploid	Two copies per cell: eukaryotes are diploid
Contain little **repetitive DNA**; most of DNA (~90%) coding	Contain large amounts of repetitive and non-coding DNA
Extrachromosomal DNA as **plasmids**	Extrachromosomal DNA in mitochondria and chloroplasts
Genes are organized into operons: clusters of co-regulated genes which produce polycistronic **mRNA**	Monocistronic mRNA is produced
Intronless genes	Genes with **introns** and **exons**

Table 6.1 A comparison of key features of prokaryotic and eukaryotic genomes

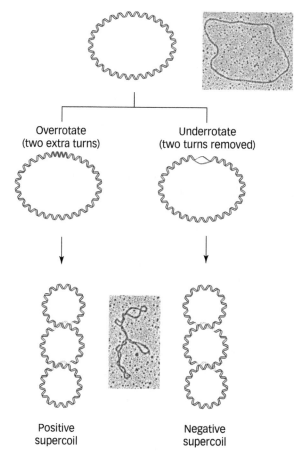

Figure 6.1 Production of supercoiled DNA, with electron micrographs of relaxed and supercoiled *Escherichia coli* DNA.

- Generally, DNA in cells is negatively supercoiled. This is advantageous for the cell as the two strands are under-rotated, which eases strand separation during replication and transcription.
- A small, basic, histone-like protein, **HU protein**, binds to DNA, inducing bends that are nicked and supercoiled by topoisomerases.

Plasmids

- Plasmids are found in most bacterial species. They are also present in **Archaea** and some lower eukaryotes, e.g. the 2u circle of the yeast *Saccharomyces cerevisiae*.
- Plasmids:
  - are small, separate, circular DNA molecules, typically 2–10,000 nucleotides in length

The bacterial genome

Plasmid type	Key genes	Examples
Fertility	**Pilus** protein for conjugation tube	**F plasmids** of *Escherichia coli*
Resistance	Antibiotic resistance	**R plasmids** of *E. coli*
Killer	Bacteriocins: toxic to other bacteria	Colicin-producing **col plasmids** of *E. coli*
Degradative	Enzymes for digestion of unusual substances	Toluene-metabolizing **Tol plasmid** of *Pseudomonas putida*
Virulence	Pathogenicity	**Ti plasmid** of *Agrobacterium tumefaciens* that induces crown gall disease in plants

Table 6.2 Types of plasmids

- ○ are present in single or multiple copies (10–20) and replicate independently using bacterial cellular enzymes
  - ○ contain only a few genes, which are related to their specific functions (Table 6.2); for other functions, such as DNA replication, they use bacterial enzymes.
- Plasmid genes can be crucial for bacterial survival in stressful situations, e.g.:
  - ○ R plasmids confer antibiotic resistance
  - ○ F plasmids give bacteria new genes that aid survival in changing environments.
- Plasmids move from one bacterium to another—even between bacteria of different species.
- Plasmids are used as cloning vectors.
  - ➲ *See section 14.1 (p. 276) for details of plasmid cloning vectors.*

F plasmids

- F or 'fertility' plasmids confer 'fertility' on bacterial cells. The term 'fertility' refers to a mechanism called conjugation (Figure 6.2) for exchanging bacterial genes.
- F plasmids can exist in two states:
  - ○ free in the bacterial cytoplasm where they replicate independently of the bacterial chromosome; the bacterial cell is termed an **F⁺ cell**
  - ○ integrated in the bacterial chromosome where they are replicated as part of the bacterial chromosomal replication process, which produces an **Hfr** bacterial cell.
- An integrated plasmid is called an **episome**.
- The two states (integrated or free in cytoplasm) are interchangeable. Sometimes, excision of an F plasmid from the bacterial chromosome is aberrant and results in bacterial genes being incorporated in the plasmid. The bacterial cell is then referred to as **F'**.

Description of bacterial cell	F plasmid
F⁺	Free in cytoplasm
Hfr	Integrated in bacterial chromosome
F'	Free in cytoplasm, containing bacterial genes
F⁻	Absent

Table 6.3 Relationship between bacterial and F plasmid genomes

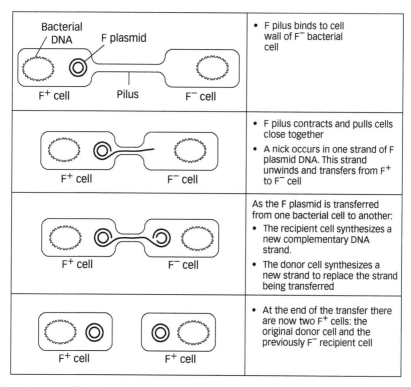

Figure 6.2 Transfer of an F plasmid via conjugation

- **F⁻ cells** lack an F plasmid.
- Most of an F plasmid's 25–40 genes code for production of a **sex pilus**.
- A pilus is a hollow, hair-like protein extension of the plasma membrane. It forms a temporary cytoplasmic bridge between two bacteria, through which an F plasmid moves from one bacterium to another.
- The process of F plasmid transfer via a pilus is called **conjugation**.
- Conjugation occurs from an F plasmid-containing bacterium (F⁺ cell) to a bacterium lacking an F plasmid (F⁻ cells) (Figure 6.2).

R plasmids

- Also termed **R factors**, these plasmids have genes that confer resistance to antibiotics such as streptomycin, tetracycline, ampicillin, sulphonamide and chloramphenicol.
- Some R plasmids possess genes for the production of a conjugation pilus. This enables R plasmids to move between bacteria, spreading antibiotic resistance.
- Bacteria with different R plasmids have multiple antibiotic resistance.
- Some exotoxins, such as the tetanus exotoxin from *Clostridium tetani* and *E. coli* shiga toxin, are also encoded by plasmids.

Isolating bacterial mutants in the laboratory

- R plasmids are used in **gene cloning**.
  - The gene of interest is inserted into a plasmid resistant to an antibiotic, usually ampicillin.
  - The recombinant plasmid is introduced into *E. coli* that are sensitive to ampicillin.
  - The bacteria are spread on an agar plate containing ampicillin. Only bacteria that have acquired the plasmid grow.
  - These plasmid-containing bacteria are cultured in ampicillin medium, replicating the gene of interest as they divide.
- In cloning, the antibiotic resistance gene is referred to as 'selectable marker': the means of selecting for cells with the recombinant plasmid.
- There are many different commercially available cloning plasmids.

  ➔ *See figure 14.1 for details of a frequently used cloning plasmid.*

Looking for extra marks?

During development of crown gall disease the Ti plasmid transfers *Agrobacterium* genes into the cells of an infected plant: part of the Ti plasmid integrates into the plant genome. This integration feature is used to genetically modify plants, for example to create 'golden rice', with genes to synthesize beta-carotene, the precursor of vitamin A.

 Check your understanding

6.1 What is a plasmid?
6.2 How does a prokaryotic and eukaryotic chromosome differ?

6.2 ISOLATING BACTERIAL MUTANTS IN THE LABORATORY

- Bacteria are either grown in liquid culture or on solid media, i.e. Petri dishes with nutrient-containing agar.
- Bacteria are grown in/on either:
  - **minimal medium**, which contains a carbon energy source (e.g. glucose), sources of nitrogen and sulphur, some inorganic ions and water
  - **complete medium**, which contains all the organic nutrients that bacteria might need.
- Wild type or **prototrophic** bacteria can grow on unsupplemented minimal medium.
- **Auxotrophic** bacteria, with a mutation in genes that encode enzymes used in essential biochemical pathways, need to be cultured on complete medium.
- An **auxotrophic mutant** can be identified by manipulating culture conditions. The technique of **replica plating** is used.
  - Bacteria are first plated on complete medium agar and allowed to form **colonies** (a clump of genetically identical bacteria derived from a single cell).

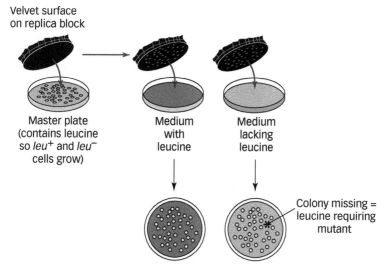

Figure 6.3 The use of replica plating to identify bacterial mutants unable to synthesize the amino acid leucine.

This usually involves diluting a liquid culture, so that plated cell numbers are low, therefore producing separated colonies (Figure 6.3).

○ A sterile velvet pad is then pressed onto the surface of this **master plate**. It picks up a few cells from each colony.

○ The pad is then pressed onto the surface of both complete and minimal media agar. Only prototrophic bacteria grow on minimal media. Any colonies not growing on the minimal medium must be auxotrophic mutants (Figure 6.3).

○ Auxotrophic mutants can be cultured from cells on the master plate and investigated further.

• The replica plating technique can be modified to select for specific auxotrophic mutants, e.g. to obtain a mutation in the gene that encodes an enzyme required for synthesis of the amino acid leucine, replica-plate bacterial cells from minimal media plus leucine to minimal medium.

• Replica plating is also useful for the identification of drug resistant strains of bacteria by replicating from plates lacking the drug (e.g. an antibiotic such as streptomycin) to those containing it. Only drug resistant colonies of bacteria would grow on the second plate.

Representation of bacterial alleles

• Three letters, indicative of a gene's function, are used to represent bacterial phenotype and genotype.

• Phenotype is indicated when first letter is capitalized (upper case); genotype when all three letters are lower case and italicized.

• If more than one gene is used in a pathway to produce a substance then the related genes are designated by letters (Table 6.4).

Gene transfer between bacteria

Nomenclature	Description
Leu	**Phenotype:** synthesis of the amino acid leucine
leu A leu B leu C leu D	**Genotype:** the four genes encoding enzymes needed for leucine synthesis
leu A$^+$ leu A$^-$	**Genotype:** functioning (wild type) allele mutant (auxotrophic) allele
Str^r/str^s	**Genotype:** streptomycin resistance/sensitivity alleles

Table 6.4 Bacterial allele nomenclature

- Superscripts:
  - +/− denote wild type/auxotrophic forms of a gene
  - s/r denote drug sensitivity or resistance.

> ### Check your understanding
>
> 6.3 With respect to bacteria, what is the difference between an auxotroph and a prototroph?

6.3 GENE TRANSFER BETWEEN BACTERIA

- **Horizontal gene transfer** is the transfer of genes between cells of the same generation by a mechanism other than reproduction. Vertical gene transfer occurs during reproduction.
- Horizontal gene transfer is important in the generation of genetic variation within a bacterial species as bacteria reproduce asexually by **binary fission**, which produces genetically identical daughter cells.
- Genetic variation is generated through recombination between transferred DNA and the recipient's genome, i.e. crossing over takes place between homologous sequences in the transferred and recipient's DNA (e.g. Figure 6.4).
- There are three gene transfer mechanisms.
  - Conjugation: plasmid-mediated gene transfer. Two bacteria lie close to each other and a connecting tube (pilus) forms through which an F plasmid carries part of a bacterial chromosome.
  - **Transformation**: a bacterium picks up DNA from its surrounding environment.
  - **Transduction**: bacteriophages act as vectors, carrying DNA from one bacterium to another.
- The amount of DNA transferred from donor to recipient is low—generally 3% or less of a donor's DNA.

- Geneticists map genes on bacterial chromosomes from information gained from recombination events following the transfer of DNA between bacterial cells.
- There is a common gene mapping strategy for all three modes of gene transfer.
  - Two strains of bacteria are used with different alleles at a number of genes. For example:
    - **recipient** strain—a⁻ b⁻ c⁻ (auxotrophic for nutrients a, b, and c)
    - **donor** strain—a⁺ b⁺ c⁺ (wild type for using/making these three nutrients).
  - The transfer of one gene relative to another is analysed by plating cells on selective media lacking the nutrients under test.
  - Recombination between transferred donor DNA and the recipient chromosome enables the recipient strain to grow on the selection media.

Conjugation

- The basic F plasmid movement through a pilus (Figure 6.2) does not involve the transfer of any bacterial genes.
- Transfer of bacterial genes occurs if conjugation occurs from an **F′** or **Hfr** cell to an F⁻ recipient bacterium (Table 6.3).
- Consider an **Hfr cell** (Figure 6.4):
  - the integrated F plasmid controls production of a pilus between the Hfr donor bacterium and the F⁻ recipient cell
  - a single strand of the integrated F plasmid is nicked and the F plasmid moves through the pilus into the F⁻ bacterium

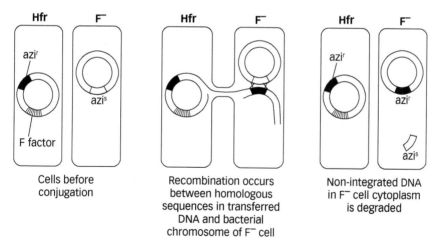

Figure 6.4 Conjugation between Hfr and F⁻ cells, and subsequent recombination between transferred bacterial DNA and F⁻ host cell DNA (for clarity, only one bacterial gene is shown (aziʳ), which denotes ability to grow on media containing sodium azide).

- because the moving F plasmid was integrated into the bacterial chromosome, the linked chromosome moves with the F plasmid
- bacterial DNA moves first and the F plasmid last.
- The amount of the bacterial chromosome that moves into the F⁻ cell depends on the length of time that the two cells remain linked by the pilus.
- Transfer of the entire *E. coli* chromosome, for example, takes 100 minutes; cells rarely remain linked for this length of time so the F⁻ bacterium remains without an F plasmid.
- However, recombination occurs between the DNA of the transferred bacterial chromosome and that of the recipient F⁻ cell (Figure 6.4).
- Fragments of non-integrated Hfr donor DNA and F plasmid remaining in the cytoplasm are digested by nucleases. Thus, the recipient cell remains F⁻, unlike conjugation between F⁺ cells and F⁻ cells (Figure 6.2), when the F⁻ cells acquire an F plasmid.

Using conjugation to map bacterial genes

- Mapping of bacterial genes:
    - uses Hfr and F⁻ cells of different genotypes (Box 6.1)
    - involves interrupting conjugation at regular intervals
    - relies upon the occurrence of recombination between transferred genes and the F⁻ genome.
- Recombination changes the genotype of the F⁻ cells. Genetically altered F⁻ cells are detected by their changed growth ability on selective culture media (Box 6.1).
- The relative position of genes on the bacterial chromosome is deduced from the time taken for different genes to be transferred from Hfr to F⁻ cells, detected by the time it takes for the expression of selected genes to change in the F⁻ cells (Box 6.1)
- Numerous Hfr strains are used when mapping a bacterial chromosome because the F factor integrates randomly and in different orientation.
- Thus, the starting point of transfer of bacterial genes during conjugation varies depending on the site of F factor integration.
- The order and relative chromosomal positions of genes are therefore inferred from the results of gene transfer between a number of different donor Hfr and recipient F⁻ strains.

Check your understanding

6.4 What is the difference between an F⁺ and Hfr strain of bacterium? How is an F⁺ strain converted into an Hfr strain?

Box 6.1 Mapping bacterial genes by interrupted conjugation—an example

The genotypes of the donor Hfr and recipient F^- strains were:

Hfr: str^s thr^+ azi^r ton^r lac^+ gal^+

F^-: str^r thr^- azi^s ton^s lac^- gal^-

($thr^{+/-}$ $lac^{+/-}$ $gal^{+/-}$ $azi^{r/s}$ $ton^{r/s}$ denotes ability/inability to grow on media lacking amino acid threonine, and containing sugars lactose and galactose or sodium azide or bacteriophage T1).

- Donor Hfr and recipient F^- strains were mixed and conjugation interrupted at regular intervals by vigorous agitation.
  - Aliquots of cells were plated on a series of selective media which selected for F^- cells that contained donor Hfr DNA.
  - The first test medium contained streptomycin and lacked threonine. Both donor and recipient bacterial strains cannot grow on this medium: only cells containing transferred Hfr DNA survive (look at genotypes and work this out).
  - Bacterial cells that grew on this medium were then plated on a series of media variously containing sodium azide, phage T1, lactose, and galactose.
- The ability to grow on these media was time related, i.e. it depended on when conjugation was disrupted. This reflects when the relevant alleles were transferred from donor Hfr to recipient F^- strains.
- The order and relative position of genes azi^r, ton^r lac^+ and gal^+ are determined by plotting the numbers of recipient F cells that have gained these selectable traits against the times when conjugation was disrupted.
- In this example recipient F^- cells became azi^r after 9 minutes, ton^r after 10 minutes, able to breakdown lactose $lacr^+$ after 15 minutes, and galactose gal^+ after 17 minutes. So we have a bacterial map:

(a) Time of gene transfer

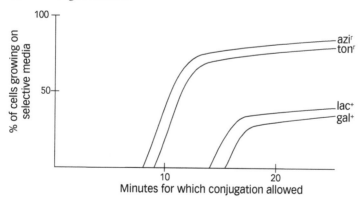

(continued)

Gene transfer between bacteria

(b) genetic map

- NB, the basic unit in bacterial maps is a minute, which is in contrast with eukaryotic gene map distances expressed as % recombination (or cM).

- Conjugation is the best way of mapping bacterial genes as it gives both:
  - the order of genes
  - the relative distances between genes.
- Transformation and transduction are used to map genes for those bacterial species in which conjugation does not occur.
- These processes only establish the order of genes because in both transformation and transduction only small segments of DNA enter a bacterial cell. By analysing which genes co-infect through recombination, maps of the order of genes are constructed.

Transformation

- Transformation is the spontaneous uptake of DNA by a bacterium from the surrounding medium, which enables DNA to be transferred from a donor to a recipient bacterium.
- Bacterial cells that take up DNA through their membrane are termed **competent.**
  - Bacteria can be naturally competent, taking up DNA when dead bacteria break up and release DNA fragments into the environment.
  - Bacteria can also be made competent by artificial means.
- Short-term exposure of bacterial cells to calcium chloride, heat shock, or an electrical field makes the cell membrane permeable to DNA.
- Artificially-induced transformation is a key step in gene cloning. It is used to introduce a recombinant plasmid into bacteria.
  ➡ *See section 14.1 (p. 274) for details of gene cloning.*
- The DNA taken up by a competent bacterium can be any type: bacterial, viral, or even eukaryotic.
- Uptake of DNA into eukaryotic cells can be similarly induced when the process is termed **transfection.**

Using transformation to map bacterial genes
- Two strains of bacteria are used with different alleles at a number of genes, e.g.:

Recipient strain:	a^- b^- c^- (auxotrophic for nutrients a, b, and c)
Donor strain:	a^+ b^+ c^+ (wild type for using/making these three nutrients)

- Donor strain DNA is purified and fragmented, and mixed with competent recipient strain cells. DNA fragments will enter the recipient bacteria and recombine with homologous DNA sequences in the recipient's chromosome (Figure 6.5).
- Recipient cells are plated on nutrient-lacking media. In this 'nutrients a/b/c' example, any cells that grow will have acquired wild type genes through transformation followed by recombination.
- **Genes are mapped** by observing the rates at which two or more genes are co-transferred, as indicated by the ability of recipient transformed cells to grow on the nutrient-lacking media.
- The relative rate at which pairs of genes are co-transformed indicates the distance between them: the higher the rate of co-transformation, the closer the genes on the bacterial chromosome, i.e. the rate of co-transformation is inversely proportional to the distances between genes.

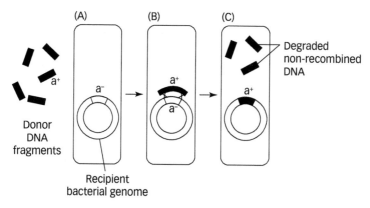

Figure 6.5 Horizontal gene transfer via transformation.

Check your understanding

6.5 How can bacteria acquire genetic diversity through the processes of conjugation and transformation?

6.4 VIRUSES: GENERAL FEATURES

- Viruses are non-cellular infectious agents that can only reproduce within a host cell: they are 'obligate intracellular parasites'.

- Outside a cell, a viral particle is called a **virion**.
- A virion consists of its genetic material encapsulated within a protein coat, known as a **capsid**.
- Some virions also become surrounded by an **envelope** of modified host membrane as they leave the host cell.
- Virions have characteristic, often highly symmetrical, shapes, e.g. helical or polyhedral forms.
- Viral genomes can be DNA or RNA.
- Most viral genomes are small, ranging in size from 1 kb to 1 Mb.
- The few viral genes typically encode:
  - the capsid protein
  - key enzymes needed for infection or replication of their genetic material.
- Viruses infect all prokaryotes and eukaryotes.
- They are host species, or even host cell type specific. The capsid protein recognizes and binds to specific receptor proteins in the host plasma membrane.
  - For example, the gp120 envelope protein of HIV-1 binds to CD4 of T helper cells and other cells of the human immune system.
- Viruses reproduce rapidly, producing large numbers of progeny. This feature, together with their small, easily sequenced genomes, has made them a favoured organism for genetic research.

  There is tremendous variability in viral life cycle details. Key features are described by considering bacteriophages and retroviruses.

6.5 BACTERIOPHAGES

- Viruses that infect bacteria are called **bacteriophages**, or simply **phages**.
- Phages commonly have a spherical icosahedral head with a tail, which may be long or short, contractile of non-contractile (Figure 6.6).
- A few phages lack a tail, e.g. ΦX174 and MS2 phages consist solely of a head.
- Most phage genomes are double-stranded DNA. For example, lambda, a popular cloning vector, and T4.
- ΦX174 is an example of a single-stranded DNA phage.
- The life cycle of a phage may be:
  - **lytic**—the host bacterial cell is killed in the production of several hundred progeny phage; the phage is described as **virulent**
  - **lysogenic**—the phage does not immediately kill the bacterial cell. Instead, its genome is incorporated into, and replicated at the same time as, the host bacterial chromosome; the phage is described as **temperate** or **non-virulent**.
- Bacteria have a defence mechanism against phages—their **restriction enzymes**, which recognize and cut up foreign DNA. Natural selection, however, favours phage mutants that are resistant to these nucleases.

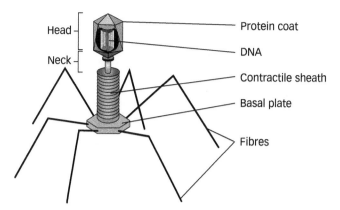

Figure 6.6 Generalized T-even bacteriophage structure.

Lytic life cycle
- The lytic cycle takes only 20–30 minutes at 37°C.
- The phage attaches to specific receptors on the bacterial cell wall and injects its DNA into the cell.
- Inside the bacterium, the phage DNA is replicated, transcribed, and translated, thus producing more phage DNA and phage protein.
- New phages are assembled.
- The phage finally produces a lytic enzyme that breaks open the host bacterium, releasing new phages (Figure 6.7).

Lysogenic life cycle
- As with the lytic cycle, the phage attaches to the bacterial cell wall and injects DNA into the cell. It then becomes inactive, integrates into the bacterial chromosome, and is known as a **prophage**.
- Prophage DNA is replicated at the same frequency as the bacterial chromosome. It is not transcribed.
- Various stimuli (e.g. nutrient depletion or ultraviolet light) cause the prophage to dissociate from the bacterial chromosome and enter into a lytic cycle, during which new phages are produced and, ultimately, the cell lyses.

Transduction
- Transduction is the transfer of genes between bacteria by bacteriophages.
- Donor bacterial DNA is packaged within the protein coat of a phage and transferred to a recipient bacteria when the phage infects it.
- In **generalized transduction** phages randomly pick up donor bacterial DNA during the lytic cycle.
    ○ Following entry of a phage into a bacterium, the bacterial DNA is digested.

Bacteriophages

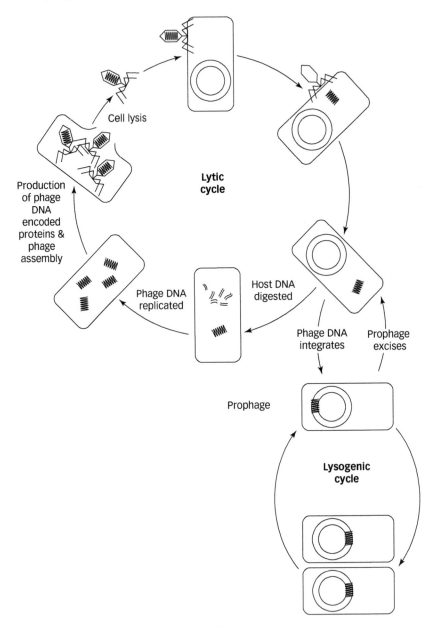

Figure 6.7 Lytic and lysogenic phage life cycles.

- At the end of the lytic cycle phage DNA is packaged into new phage particles.
- A fragment of the host bacterial DNA, which was digested when the original phage entered the cell, is occasionally packaged into a phage coat instead of phage DNA. This produces a **transducing phage**.

- When a transducing phage enters a new cell, the packaged bacterial DNA is released and may recombine with the bacterial chromosome. In this way bacterial genes can be moved from one strain to another, but with very low frequency. Recombinant bacteria are termed **transductants**.
- The fragment of bacterial DNA transferred into the recipient bacterial cell is small. Thus, only genes close together will be co-transduced.
- Rates of gene co-transduction, like co-transformation rates, can be used to **map** bacterial genes. Co-transduction gives an indication of order and physical distance of genes on a bacterial chromosome.
  - As with mapping by transformation, two bacterial strains with different alleles at several loci are used. The donor strain could be $a^+ b^+ c^+$ which is infected with the phages. The progeny phages are then mixed with recipient bacteria of genotype $a^- b^- c^-$ and plated on different media to determine the phenotypes and therefore genotypes of transduced bacteria.
- **Specialized transduction** is linked to a lysogenic life cycle when phage DNA integrates into a bacterial chromosome.
- Excision of the prophage from the bacterial DNA may be imperfect. A small fragment of bacterial DNA may also be linked to the prophage.
- Integration occurs at specific sites in the bacterial chromosome. Thus, during specialized transduction, the prophage will only transduce bacterial genes close to the integration site.
- Specialized transduction only yields mapping information of the few genes close to the phage DNA integration site.

> ### Check your understanding
>
> 6.6 What is the difference between a lytic and a lysogenic life cycle?

6.6 EUKARYOTIC VIRUSES

- All eukaryotic species are susceptible to viral infections.
- Most plant viruses have RNA genomes, e.g. tobacco mosaic virus.
- In contrast, the genomes of animal viruses are highly variable. They can be:
  - DNA or RNA
  - double- or single-stranded
  - circular or linear
  - continuous or segmented.
- Many RNA and single-stranded DNA viruses have a **segmented** genome, i.e. it is split into smaller, often single gene, pieces, e.g. the RNA genome of influenza virus.
- Segmented genomes are believed to be a strategy to reduce error rates during genome replication.

Eukaryotic viruses

- Segmented genomes produce viral variability. Different strains of a virus with a segmented genome can shuffle and combine genes (alleles) thus producing progeny viruses with unique characteristics.
- **DNA viruses** are grouped into those with:
  - a double-stranded DNA genome, e.g. smallpox virus
  - a single-stranded DNA genome, e.g. adeno-associated virus.
- There are four main groups of **RNA viruses** (Table 6.5). RNA viruses include many medically important human viruses, e.g. influenza, cold, and AIDS viruses
- Virus genomes range in size from the 2 kb nucleotides of circoviruses, which codes for just two proteins, to ~1.2 million nucleotides of mimiviruses, coding for a thousand proteins.
- RNA genomes are generally smaller, comprising just a few genes. The lack of an RNA replication error correction mechanism is believed to limit RNA genome size.
- The few genes of an RNA virus are used in a highly efficient manner (see the next section on retroviruses).

Retroviruses

- The life cycle of a eukaryotic virus is illustrated by that of **retroviruses**.
- Retroviruses are single-stranded (positive sense) RNA viruses which replicate their genome via a DNA copy.
- Typically, a retrovirus's RNA genome is 7–12 kb in length, containing 3 genes.

RNA genome	Examples	Description
Single-stranded positive (+) sense	Poliovirus Rhinovirus Tobacco mosaic virus	• RNA genome has 'message sense' • It can function as mRNA and be translated immediately upon virion infection of a host cell
Single-stranded negative (−) sense*	Influenza Measles Rabies	• RNA genome is complementary to the 'message sense' • It is copied into complementary positive-sense mRNA before **translation** by the **RNA replicase**, i.e. RNA-dependent RNA polymerase, which is packaged in the virion ready for use upon cell infection
Double-stranded	Rotaviruses	• A double-stranded RNA genome cannot function as mRNA • Viral encoded RNA polymerase makes mRNA following virion infection
Single-stranded positive (+) sense: retroviruses	HIV-1 Rous sarcoma	• Although the RNA genome has 'message sense', it cannot function as mRNA immediately upon infection • Genomic RNA is reverse transcribed into DNA, which is then transcribed into functional mRNA

Table 6.5 The various genomes of RNA viruses

*Ambisense genomes have a mixture of positive and negative sense molecules.

Retrovirus genes

- Three genes (*gag, pol,* and *env*) are common to all retroviruses.
  - *Gag* encodes viral coat proteins.
  - *Pol* encodes a reverse transcriptase and integrase.
  - *Env* encodes the viral surface glycoprotein and transmembrane proteins that mediate cellular receptor binding and membrane fusion.
- Multiple proteins are produced from these three genes depending upon processing by the protease product of a fourth gene, the *pro* gene.
- The *pro* gene sequence is at the beginning of the *pol* gene. Sometimes it begins earlier, in the *gag* gene (Figure 6.8).
- The gag, pol, and pro proteins are translated from full length genomic RNA.
- The env protein is expressed from mRNA that is spliced using host cellular apparatus.
- A fourth gene, an **oncogene**, is present in tumour-forming retroviruses.
- The *src* oncogene encodes a tyrosine kinase that attaches phosphate groups to the amino acid tyrosine in host cell proteins. This triggers changes in many fundamental cellular processes, e.g. stimulates uncontrolled mitosis.

Retrovirus life cycle

- When a retrovirus is injected into a host cell its RNA genome is copied into DNA, which integrates into the host cell's genome.
- Once inside the cell a **reverse transcriptase**, packaged in the virion, synthesizes a DNA molecule from the viral RNA in a three-stage process.
  - Its **RNA-dependent DNA polymerase** domain transcribes a complementary DNA strand from the viral RNA, forming an RNA/DNA hybrid. A tRNA captured from the host cytoplasm primes this synthesis.
  - Its **RNase H** activity degrades the template RNA and primer tRNA.
  - The **DNA-dependent** DNA polymerase domain synthesizes a DNA strand complementary to the DNA strand.
- The newly synthesized DNA molecule is inserted into a host cell chromosome by the viral **integrase**. The integrated viral genome is termed a **provirus**.

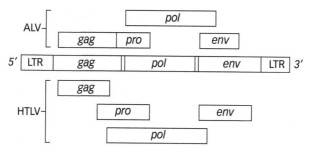

Figure 6.8 Generalized retrovirus genome showing transcription products in avian leukosis virus (ALV) and human T-lymphotropic virus (HTLV). LTR: long terminal repeats.

- The provirus is replicated by host enzymes whenever the host chromosome is duplicated.
- At an appropriate time (days to years following integration), the provirus is transcribed. Multiple copies of the viral RNA genome are produced which is both:
  - translated, producing viral proteins
  - packaged into new viral particles as genomic RNA.
- As these particles are released from the host cell they are enveloped in fragments of the cell membrane.
- Following transcription of the provirus, retroviral RNA transcripts are subject to the same processing events as eukaryotic mRNAs, including **cap** addition at the 5′ end, polyadenylation at the 3′ end, and splicing to form subgenomic-sized RNA molecules.

Check your understanding

6.7 The base composition of a virus was found to be 13% A, 30% G, 19% U, and 38% C. Is this a DNA or RNA virus? Is it single- or double-stranded?

6.8 How do retroviruses, such as HIV, infect cells and integrate their genomes into the host chromosome?

6.9 Identify similarities and differences in HIV and phage structure and cellular infection.

6.7 ORGANELLE GENETICS

- Mitochondria and chloroplasts differ from other eukaryotic organelles as they:
  - contain DNA
  - are only produced by growth and division of pre-existing organelles.
- Mitochondria and chloroplast DNA (mtDNA and cpDNA) is more similar to bacterial than eukaryotic DNA:
  - a small, circular (supercoiled) chromosome lacking histones
  - present in the organelle as condensed nucleoid structures
  - little repetitive DNA
  - intron-less genes
  - its mRNA is translated by a bacterial style system
- The bacterial nature of mitochondrial and chloroplast genomes is explained by the **endosymbiotic theory**:
  - that free-living bacteria were engulfed by primitive nucleated cells; host and bacteria established a symbiotic relationship
  - during subsequent evolution, many bacterial genes were lost or transferred to the nuclear genome, leaving a reduced-sized genome.

- The assembly and function of mitochondria and chloroplasts require gene products from both the organelle and nuclear genomes.
- Organelle genomes encode the rRNA genes and most of the tRNA genes of the organelle translation system, and varying numbers of functional protein genes.

Mitochondrial and chloroplast genomes

- Organelle genomes are small (Table 6.6). For example, the human mitochondrial genome consists of 16,569 nucleotides which encode 37 genes.
- mtDNA varies considerably in size between different species, from 6 kb in the *Plasmodium* parasite to 240 kb in muskmelons.
- Chloroplast genomes tend to be larger, but more uniform, in size (120–180 kb).
- cpDNA encodes more genes than mtDNA:
  - cpDNA encodes ~100 polypeptides—components of the light-dependent and light-independent stages of photosynthesis, RNA polymerase, translation factors, ribosomal proteins, and other chloroplast factors
  - mtDNA encodes 10–20 polypeptides of the electron transport chain used in oxidative phosphorylation (out of a total of 80+ polypeptides).
- The mitochondrial genome size differences do not always reflect comparable differences in gene content. For example, the yeast 78 kb mtDNA encodes fewer genes than the human 16.5 kb mitochondrial genome, which is particularly compact (Table 6.7).
- The mitochondrial genome is also more variable in structure and organization. It consists of:
  - a single, double-stranded circular DNA in animals and fungi

Genome	Gene number	C value (kb)
Mitochondrion/chloroplast	5–100	6–240
Bacterial nuclear genome	500–10^4	500–10^4
Eukaryotic nuclear genome	10^4–10^5	10^4–10^8

Table 6.6 Comparison of organelle, prokaryotic, and eukaryotic genome sizes

	Human	Yeast
Genome size (kb)	16.5	78
Coded genes	2 rRNA 22 tRNAs 13 polypeptides	2 rRNA 25 tRNAs 7 polypeptides
Non-coding intergenic DNA	• Adjacent genes abut or overlap slightly • D loop (which contains origin of replication) only non-coding region	Many long intergenic sequences
Introns	Absent	Present
Gene mutations	Often affect muscles and the eye, e.g. LHON	Petite

Table 6.7 Comparison of human and yeast (*Saccharomyces cerevisiae*) mitochondrial DNA
LHON: Leber's hereditary optic neuropathy.

- o multiple circular DNA molecules in many plant species
- o a large interlocking network of mini- and maxi-interlocking DNA circular molecules in Protoctista parasitic genera, e.g. *Trypanosoma*
- o a linear chromosome in the ciliated Protozoa, *Paramecium*, alga *Chlamydomonas,* and yeast *Hansenula.*
- Generally, the chloroplast genome is a linear chromosome.
- Each eukaryotic cell contains multiple copies of its organelle DNA because there are:
  - o many mitochondria and chloroplasts (up to 10^3) per cell
  - o many copies of its genome per organelle (2–10/organelle).
- Mitochondrial DNA typically accounts for 1% of total cellular DNA; levels of cpDNA may reach 10% in young green shoots.
- These high cellular levels of mtDNA and cpDNA have led to organelle genes being used for the Barcode of Life Project (see section 14.4, p. 292).
- Inheritance of organelle genes does not follow Mendelian principles because mitochondria and chloroplasts are inherited maternally (see section 3.5 and Figure 6.10) via the egg cytoplasm.
- Furthermore, expression of different alleles of a gene in an individual is highly variable because of the high copy number of mtDNA and cpDNA per cell.
- **Homoplasmic** cells or organisms have one allele for an organelle gene.
- **Heteroplasmic** cells or organisms have two or more alleles for an organelle gene
- Alleles in different heteroplasmic cells are often in different proportions because of the following three factors:
  - o organelle genomes are selected randomly for replication
  - o they are partitioned randomly when an organelle divides
  - o organelles are segregated randomly when a cell divides.
- These stochastic mechanisms produce different frequencies of a mutant allele in different cells and tissues (Figure 6.9), and explain the extreme variability in the expression of inherited mitochondrial and chloroplast mutations.
- Replication, transcription, and translation of mtDNA and cpDNA have many characters distinctive from eukaryotic processes.
  - o mtDNA and cpDNA are replicated throughout the cell cycle.
  - o mtDNA and cpDNA molecules are chosen randomly for replication.
  - o DNA polymerase γ is used, which lacks proofreading so there is a high mutation rate of mtDNA and cpDNA.
  - o The **genetic code** of mitochondria differs from the universal code of nuclear DNA (see section 9.1, p. 172).
  - o RNA editing of pre-mRNA occurs frequently. For example, some plant species add or delete **cytosine**. Trypanosomes add and delete various **uracils**.
  - o Many aspects of mitochondrial and chloroplast translation resemble bacterial rather than eukaryotic protein synthesis. For example:
    - ▪ bacterial-sized **ribosomes** (70S) and rRNA (16S and 23S)

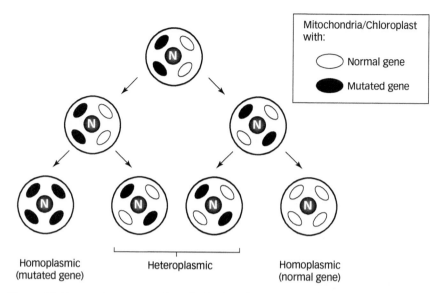

Figure 6.9 The random segregation of mitochondria/chloroplasts during cell division. N: nucleus.

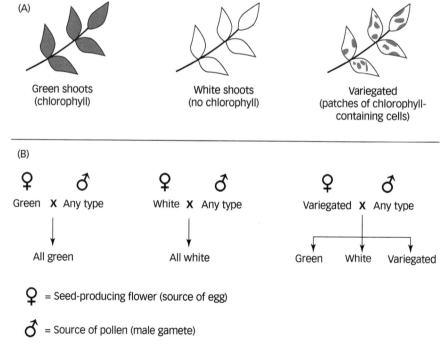

Figure 6.10 Non-Mendelian inheritance patterns of organelle DNA: inheritance of leaf variegation in the four-o'clock plant *Mirabilis jalapa*. (A) Different chlorophyll distribution patterns; (B) chlorophyll inheritance patterns.

- tRNA met, carrying formyl-methionine, initiates translation
  - it is sensitive to inhibitors of bacterial translation, such as chloramphenicol and streptomycin.
- mtDNA and cpDNA have high mutation rates owing to lack of organelle DNA repair mechanisms.

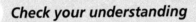

Looking for extra marks?

Recent research has shown that, among mammals, a species maximum longevity correlates well with mitochondrial genomic GC content, i.e. species with a CG-rich genome tend to live longer. It is presumed to be an adaptive trait, stabilizing and protecting the mitochondrial genome from mutations.

 Check your understanding

6.10 To what extent are mitochondria and chloroplasts dependent upon the nuclear genome?

6.11 Is your mitochondrial DNA inherited from your maternal or paternal grandmother?

6.12 What is the difference between chloroplast and Mendelian inheritance?

online resource centre You'll find guidance on answering the questions posed in this chapter—plus additional multiple-choice questions—in the Online Resource Centre accompanying this revision guide. Go to **www. oxfordtextbooks.co.uk/orc/thrive/** or scan this image:

7 The Biochemical Basis of Heredity

DNA is the universal genetic material with the exception of some viruses that use RNA.

Key concepts

- Nucleotides are the building blocks of DNA and RNA.
- Nucleotides polymerize to form long chains.
- A DNA molecule consists of two chains of nucleotides coiled around each other to form a double helix.
- An RNA molecule consists of a single chain.
- The DNA double helix is formed by complementary base pairing between nucleotides on opposite strands.
- During DNA replication the two strands separate and each is a template for synthesis of a new strand.
- The sequence of nucleotides encodes the genetic information.
- The genome of many eukaryotic species contains a high percentage of repetitive, non-coding DNA.
- Transposable elements make up the bulk of repetitive DNA.

7.1 PRIMARY STRUCTURE OF DNA

The primary structure of DNA consists of a linear sequence of nucleotides linked by phosphodiester bonds.

Nucleotides

- A nucleotide consists of a sugar, a phosphate, and a **nitrogenous base**.
- The sugar is a five-carbon or **pentose** sugar.
- The pentose sugars in DNA and RNA are slightly different (Figure 7.1):
  - in DNA the pentose sugar is **deoxyribose**, with a hydrogen atom (-H) attached to the 2'-carbon atom
  - in RNA it is **ribose**, with a hydroxyl group (-OH) attached to the 2'-carbon atom.
- The phosphate consists of a phosphorus atom linked to four oxygen atoms. It bonds to the 5'-carbon atom of the pentose sugar.
- It is the phosphate group which makes DNA and RNA acidic with a negative charge.

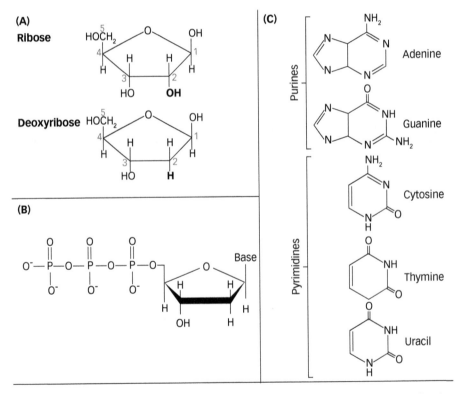

Figure 7.1 Details of nucleotide structure: (A) ribose and deoxyribose, (B) a generalized nucleotide, (C) nitrogenous bases.

- There are two types of nitrogenous base (Figure 7.1):
  - **purines**—**adenine** (A) and **guanine** (G)—which have a double ring structure
  - **pyrimidines**—**thymine** (T), cytosine (C), and uracil (U)—which have a single ring.
- Purines and pyrimidines link to the 1′ carbon of the pentose sugar by a β-N-glycosidic bond.
- The term **nucleoside** refers to the chemical unit of base and sugar.
- One or more phosphates added to a nucleoside produces a **nucleotide**, which can also be described as a nucleoside mono-/di-/triphosphate, depending on the number of phosphates attached.
- Nucleoside triphosphates (dNTPs) polymerize to produce DNA and RNA. During polymerization two phosphates are removed.
- Adenosine triphosphate (ATP) and guanosine triphosphate (GTP) are important cellular bio-energetic molecules.

Structure of a polynucleotide strand

- DNA consists of two polynucleotide strands.
- The sugar and phosphate groups are positioned on the outside, while the bases project into the interior.
- **Phosphodiester bonds** link adjacent nucleotides in each strand.
- The link is **5′ to 3′**, i.e. a phosphate bonded to the 5′ carbon of one sugar links to a free 3-OH group of a sugar in an adjacent nucleotide (Figure 7.2).

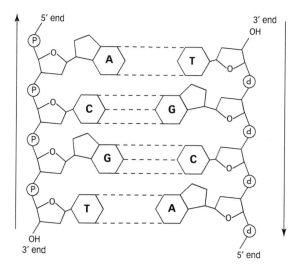

Figure 7.2 Summary of DNA structure, emphasizing the antiparallel nature of the two strands and complementary base pairing. (P: phosphates; A, G, C and T: the four bases.)

Primary structure of DNA

- A polynucleotide strand has **directionality** (Figure 7.2):
  - at the **5′ end** of a strand there is a free phosphate group attached to the 5′ carbon of the sugar
  - at the **3′ end** there is a free -OH group attached to the 3′ carbon of the sugar.
- In DNA the two polynucleotide strands are **antiparallel**, i.e. are orientated in opposite directions. One strand is orientated 3′ → 5′, the other strand 5′ → 3′. This means that the 5′ end of one strand is opposite the 3′ end of the other strand (Figure 7.2).
- The nitrogenous bases of opposite chains are paired to one another by hydrogen bonds.
- Only A=T and G≡C bonding occurs.
- A and T are held by two hydrogen bonds, and C and G by three.
- This specific A/T and C/G base pairing is referred to as **complementary base pairing**.
- Complementary base pairing ensures efficient and accurate DNA replication.
- ➜ *See section 7.8 (p. 140) for details of DNA replication.*

DNA methylation

- Methyl groups (-CH$_3$) are frequently added to bases.
- Adenine and cytosine are commonly methylated in prokaryotes.
- This methylation distinguishes bacterial DNA from foreign unmethylated DNA introduced by bacteriophages.
- Restriction enzymes recognize and cut up unmethylated DNA.
- Cytosine is methylated in eukaryotes, forming **5-methylcytosine** (Figure 7.3).
- 5-Methylcytosine occurs predominantly at CpG dinucleotides.
- DNA methylation is an epigenetic modification of DNA formed by the action of DNA methyltransferase.
- Typically, methylated sequences show lower levels of transcription.

Looking for extra marks?

DNA's methylation pattern can be identified by the technique of bisulphite sequencing. Bisulphite converts cytosine to uracil (replicated as T during the sequencing reaction), while methylcytosine is unaffected. The relative levels of C and T are analysed in bisulphite-treated and bisulphite-untreated DNA sequence.

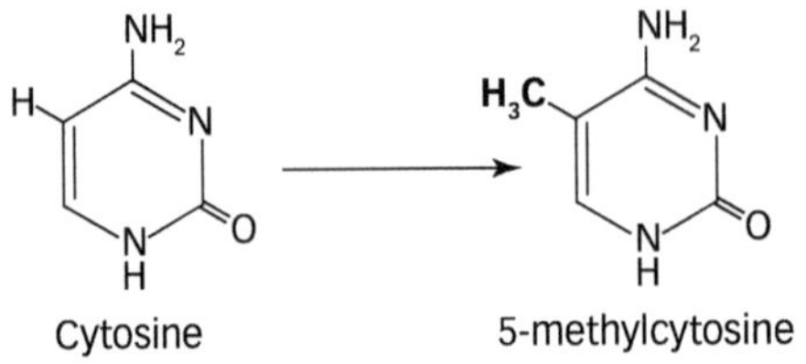

Figure 7.3 Formation of methylcytosine.

7.2 SECONDARY STRUCTURE OF DNA

- The two polynucleotide strands wind around each other to form the **double helix**. This represents the **secondary structure of DNA**.
- In nature, DNA can form three different helical structures: **A-**, **B-**, and **Z-DNA**, depending upon the conditions in which DNA is placed.

B-DNA structure

- The B-form predominates in cells, as it is the form of DNA when surrounded by plenty of water and when there are no unusual bases present.
- It is a right-handed (or alpha) helix of diameter 23.7 Å (2.37 nm).
- B-DNA makes one complete turn around its axis every 10.4 to 10.5 base pairs or 34 Å (3.4 nm).
- This means each base pair rotates ~36° relative to adjacent bases.
- The spiralling of the nucleotide strand creates alternating spiral grooves: larger **major grooves** and smaller **minor grooves**. These are important for the binding of proteins that regulate gene expression (Figure 7.4A).

A- and Z-DNA

- The **A-form** occurs in dehydrated samples, and possibly in DNA–RNA hybrids and double-stranded RNA.
- The A-form is a compacted right-handed helix with bases tilted more steeply. It is shorter and wider than the B-form (Figure 7.4B).

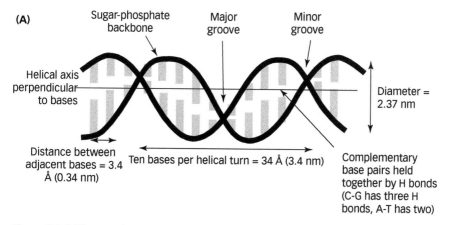

Figure 7.4 DNA secondary structures. (A) Diagrammatic representation of B-DNA showing dimensions. (B) The three forms of DNA.

Primary structure of RNA

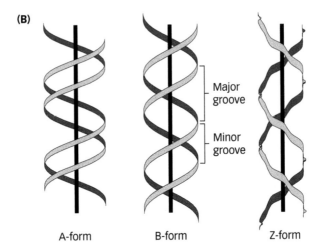

A-form B-form Z-form

Figure 7.4 (Continued)

- The **Z-form** is found in areas of DNA where:
  - DNA is being transcribed and is believed to play a role in regulation of gene expression
  - there are distinctive base sequences, such as stretches of alternating C and G.
- Z-DNA forms a left-handed helix (Figure 7.4B).

7.3 PRIMARY STRUCTURE OF RNA

- RNA consists of a single polynucleotide strand.
- There are slight differences in RNA and DNA primary structures (Table 7.1).
- RNA has many different secondary forms relating to its multiple functions.

⮕ *See section 8.3 (p. 153) for details of RNA secondary structure.*

Feature	RNA	DNA
Pentose sugar	Ribose (hydroxyl on 2′ carbon)	Deoxyribose (hydrogen on 2′ carbon)
Nitrogenous bases	A, C, G, and uracil (U)	A, C, G, and T
Molecule size	Short	Long
Polynucleotide strands	Single	Double

Table 7.1 Key differences between RNA and DNA structure

 Check your understanding

7.1 Draw a short section of RNA and DNA: highlight differences. What aspects of DNA's structure contributes to its much greater stability than RNA?

7.2 Describe the three-dimensional structure of DNA.

7.4 KEY EVIDENCE THAT DNA IS THE GENETIC MATERIAL

- DNA was discovered by Friedrich Miescher in 1869. He isolated DNA from white blood cells in pus on surgical bandages.
- Key experiments from the 1920s to the 1950s identified DNA as the genetic material.

Transformation experiments

- Frederick Griffiths and colleagues (1928) experimented with two strains of the bacterium *Streptococcus pneumoniae*.
- Mice died if injected with the virulent, non-capsulated strain, but lived if they received either capsulated or dead virulent bacteria.
- Furthermore, if harmless capsulated bacteria were cultured with dead virulent cells they became virulent and killed mice.
- Griffiths concluded that some part of the dead virulent bacteria had **transformed** the avirulent strain into a disease-causing strain.
- Oswald Avery, Colin Macleod and Maclyn McCarty (1944) showed that DNA was the transforming agent.
- They tested separately bacterial extracted protein, carbohydrates, RNA, and DNA for their ability to transform avirulent strains.

Radioactive labelling experiments

- Alfred Hershey and Martha Chase (1952) worked with T2 bacteriophage.
- They had two cultures of T2:
  - one grown with ^{32}P so all viral DNA was radioactive
  - the other grown with ^{35}S to radioactively label viral protein.
- New T2 phage produced by bacteria contained only ^{32}P, showing that DNA was used by viruses to pass on their genetic information to the next generation.
  ➤ *See section 6.5 (p. 114) for details of bacteriophage structure and life cycle.*

Model building

- In 1953 James Watson and Francis Crick built a plausible model of the structure of DNA.
- They gathered information from various sources:
  - their own X-ray crystallography photograph of DNA, which showed the bases positioned inside the molecule
  - Rosalind Franklin's X-ray diffraction patterns, which indicated that DNA was a double helix

- ○ Erwin Chargaff's analysis of DNA's base ratios, i.e. A+G=T+C, and amounts of A=T and of G=C.
- Their DNA model suggested a mechanism of accurate replication.

7.5 GENOMES

- The term **genome** refers to a species' haploid DNA content.
- There is huge variation among living species in the amount of DNA per genome.
- Most eukaryotic genomes are significantly larger than prokaryotic ones.
- Eukaryotic complexity relative to prokaryotes requires more genes. However, this explanation only partially explains eukaryotes' greater genome size.
- Eukaryotic genomes contain a lot of non-coding DNA; some of this DNA regulates gene expression, but most has no known function.
- Table 7.2 shows genome sizes for a range of prokaryotic and eukaryotic species.
- A species' genome size is also referred to as its **C-value**.
- The size of a genome (or the C-value) is generally stated as a certain number of base pairs (bp).
- The lack of a consistent relationship between the C-value and the complexity of an organism (compare salamanders and humans, or the ferns and the mustard in Table 7.2) is called the **C-value paradox**.
- An increasing number of species' genomes are being sequenced.

Species	Genome size (bp)	Number of genes
Φ174 (virus of *Escherichia coli*)	5386	11
λ bacteriophage	50,000	485
Mycoplasma genitalium (bacterium)	580,000	521
Mimivirus (virus of amoeba)	1,200,000	1262
E. coli (bacterium)	4,600,000	4377
Saccharomyces cerevisiae (bakers' yeast)	12,000,000	5770
Arabidiopsis thaliana (a mustard)	125,000,000	~27,000
Drosophila melanogaster (insect)	170,000,000	~13,760
Homo sapiens (human)	3,200,000,000	~21,000
Oryza sativa (rice)	4,200,000,000	~55,000
Amphiuma (salamanders)	760,000,000,000	Unknown
Psilotum nudum (fern)	2,500,000,000,000	Unknown

Table 7.2 Genome sizes (approximate) of various viral, prokaryote and eukaryote species

Check your understanding

7.3 How many copies of the genome are present in a diploid cell?

7.6 TYPES OF DNA SEQUENCE

Eukaryotic genomes possess many different types of DNA sequence.

Unique sequence DNA

- Unique sequence DNA is present once or repeated a few times in a genome.
- It includes both coding (i.e. protein-encoding) and non-coding DNA regions.
- The coding DNA includes single copy genes and gene families.
- Twenty-five to 50% of protein-encoding genes are present in single copies.
- Many other genes are present in similar, but not identical, copies referred to as **gene families**.
- Gene families arise through duplication of an ancestral gene. The gene copies have diverged in sequence, but generally encode related functions.
- Most gene families include just a few genes, e.g. the human β-globulin gene family on chromosome 11. Five functional genes are expressed at different stages of development (Figure 7.5).
- Vertebrate immunoglobulin gene families, exceptionally, contain hundreds of related sequences.
- Gene families also contain **pseudogenes**, non-functional genes with sequence homology to known structural genes, but no longer capable of being transcribed or translated.
- There are three main types defined according to their different origins.
  1. **Duplicated pseudogenes** arose through duplication and divergence of an ancestral gene. The β-globulin gene family contains two such pseudogenes (Figure 7.5).
  2. **Disabled pseudogenes** are the result of a mutation in an ancestral single copy functional gene. In all mammals except guinea pigs, bats and primates there is a functional gulon-γ-lactone oxidase (GULO) gene enabling synthesis of ascorbic acid (vitamin C). In primates, guinea pigs, and bats this function is absent as the GULO gene is only present as a disabled pseudogene.
  3. **Retro-pseudogenes** arise by insertion of a cDNA made by a reverse transcriptase of mRNA transcript of a protein-encoding gene, e.g. ribosomal protein pseudogenes in mammalian genomes.

Figure 7.5 Human β-globulin gene family on chromosome 11 (at 11p15.5). Haemoglobin is formed from two β-globin and two α-globin polypeptides together with four haem groups. The nature of the β-globin changes during development in relation to the strength of oxygen binding required. The ε gene is expressed in the embryonic yolk sac, Ay and Gy during fetal development. Adult haemoglobin uses the product of the β(mostly) and δ-genes. Ψβ1 and Ψβ2 are pseudogenes.

Types of DNA sequence

Moderately repetitive DNA

- This DNA consists of sequences that are 150–1000s of nucleotides in length and present in tens to thousands of copies scattered randomly throughout a eukaryotic genome.
- It includes both coding and non-coding DNA.
- Multi-copy coding sequences are functional genes whose products are in high demand by cells. These include genes for ribosomal and transfer RNA, and for histone proteins.
- For example, human rRNA gene clusters (each with 30–40 repeats) are found on the p arm of acrocentric chromosomes 13, 14 ,15, 21, and 22.
- Non-coding moderately repetitive DNA is also termed **interspersed repeat sequences**. They are classified into two types depending upon the length of the repeating sequence.
  1. **SINEs** (short interspersed elements) are normally a few hundred nucleotides in length. An example is the 280-bp *Alu* sequence that is present more than a million times in the human genome, of which it comprises ~11%.
  2. **LINEs** (long interspersed elements) are several thousand nucleotides long. Seventeen per cent of the human genome comprises repeats of the 6 kb LINE1.
- Interspersed repeat sequences are **transposable elements** (see section 7.6).

Highly repetitive DNA

- This DNA consists of a short sequence (200 bp or less) that is **tandemly repeated**, i.e. present in the genome as an array of consecutive repeats.
- This DNA is also called **satellite DNA** as it bands separately from the bulk DNA during density gradient centrifugation.
- There are three classes of satellite DNA.
  1. **Macrosatellite** DNA found in centromeres. For example, the **alphoid DNA** located at the centromere of human chromosomes. Its repeat unit is 171 bp and the repetitive region accounts for 3–5% of the DNA in each chromosome.
  2. **Minisatellites**, also known as **VNTRs** (variable number tandem repeats). The repeating unit is 10–100 nucleotides.
  3. **Microsatellites**, also known as **STR** sequences (short tandem repeats). The repeating sequence is very short, generally 2–4 nucleotides.

Microsatellites

- Microsatellites are used extensively in molecular biology as **molecular markers**, i.e. segments of DNA with variable forms whose inheritance can be followed.
- For example, forensic **DNA fingerprinting** techniques use variation in sizes of a suite of microsatellites.

Parental genotypes	$(CGA)_6(CGA)_{16}$	$(CGA)_{21}(CGA)_{21}$
Possible gametes	$(CGA)_6$ $(CGA)_{16}$	$(CGA)_{21}$
Children	$(CGA)_6(CGA)_{21}$	$(CGA)_{16}(CGA)_{21}$

Figure 7.6 Mendelian inheritance of microsatellites.

- A microsatellite's particular repeat unit, e.g. CGA, is referred to as a **motif**.
- The number of motif repeats produces the microsatellite length variation between individuals.
- For example, one individual might have repeats $(CGA)_6(CGA)_{16}$, while another may have $(CGA)_{21}(CGA)_{21}$. NB, each individual has two copies of each microsatellite, one on each of a pair of homologous chromosomes.
- Microsatellites show Mendelian inheritance patterns. If the two aforementioned individuals are male and female, their children would have heterozygous microsatellite genotypes $(CGA)_6(CGA)_{21}$ or $(CGA)_{16}(CGA)_{21}$ (Figure 7.6).
- Occasionally, microsatellites are found within coding regions.
- The repeating motif is a trinucleotide, or codon, encoding an amino acid.
- **Trinucleotide repeat disorders** result when the number of microsatellite repeats exceeds the normal stable threshold number.
- For example, the inherited human neurodegenerative condition of Huntington's disease is expressed when a particular CAG triplet exceeds a threshold of 35–40 repeats.
- The nucleotide sequence of most telomeres is a microsatellite. The telomere repeat of vertebrates is TTAGGG.

7.7 TRANSPOSABLE ELEMENTS

- Both prokaryotic and eukaroyotic genomes contain transposable elements: DNA segments that are capable of self-generated movement (**transposition**) from one site in the genome to another location.
- During their movement many transposable elements leave behind copies, hence building up the bulk of repetitive DNA.
- Thus, a large percentage of repetitive DNA sequences is transposable elements—also referred to as **jumping genes**, mobile elements, cassettes, and transposons.
- Transposable elements make up a large fraction of the C-value of eukaryotic DNA (section 7.5).
- **Junk DNA** largely comprises transposable elements.
- Transposable elements cause mutations by inserting into genes or by promoting chromosomal rearrangements, such as deletions, insertions, or **translocations**.

Structure of transposable elements

- Most transposable elements share two common characteristics (Figure 7.7A):
    - **terminal inverted repeats**, which are 9–40 bp in length and are recognized by enzymes that catalyse transposition
    - short direct repeats that flank each end of a transposable element—they are generated by the insertion mechanism (Figure 7.8).
- These two adjacent sets of repeats flag up the presence of a transposable element in a genome.
- The intervening sequence between the repeats may be short, encoding only those proteins directly involved in transposition. In other cases it includes genes for functions unrelated to transposition, e.g. antibiotic resistance (Figure 7.7B).
- Thus, transposable elements vary in size from as little as 50 bp to 10 kb.

Looking for extra marks?

A few bacteriophages reproduce by transposition, for example Mu. They use transposition to insert themselves into a bacterial genome during a lysogenic cycle.

Movement of transposable elements

- Depending on the sequence nature of a transposable element its movement may occur by either (Figure 7.8):
    - 'copy and paste' (**class I transposons**)
    - 'cut and paste' (**class II transposons**).

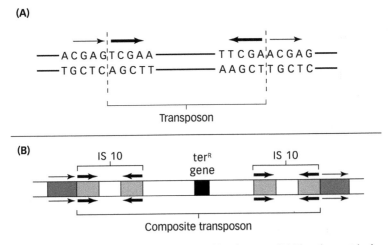

Figure 7.7 Generalized structure of a transposable element. (A) The element is shown *in situ* with adjacent direct repeats (direction of repeats indicated by thin arrows). The **inverted repeats** at the two ends of a transposon are indicated by thick arrows. (B) An example of a bacterial composite transposon: Tn10 (~9300 bp). It contains a gene for tetracycline resistance (*ter*^{R}) sandwiched between two simple transposons, IS10.

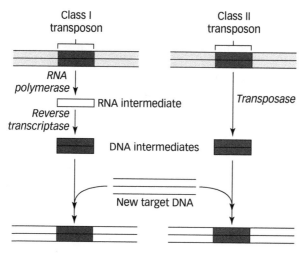

Figure 7.8 Principles of class I and class II transposition.

Class I transposons

- These first transcribe their DNA into RNA using a cellular RNA polymerase.
- A DNA copy of this RNA is then made by a reverse transcriptase and inserted into a new location in the genome.
- The reverse transcriptase is generally encoded by the transposable element.
- As these class I transposons behave very similarly to retroviruses, they are also known as **retrotransposons**.

  ➔ *See section 6.6 (p. 118) for details of retroviruses.*

- Retroviruses and retrotransposons are believed to share a common ancestor. Their key difference is that retrotransposons lack an envelope glycoprotein gene and so cannot form infectious particles.
- Retrotransposons are particularly abundant in plant genomes, e.g. they comprise ~70% of the wheat genome, in humans 42%
- The class I transposons are divided into two subtypes: long terminal repeat (LTR) and non-LTR retrotransposons. The LTR retrotransposons have direct terminal repeats ranging from ~100 to 5000+ bps.
- LINEs and SINEs (section 7.5) are non-LTR class I transposons.
  - LINEs use RNA polymerase II to produce an RNA copy, which is reverse transcribed by a LINE-encoded reverse transcriptase. They also encode an endonuclease to cut DNA for insertion of the DNA copy.
  - SINEs are transcribed by RNA polymerase II and do not encode a reverse transcriptase or endonuclease. They rely on LINE-encoded enzymes. Hence, they are also termed **non-autonomous** class I transposons.

> ### *Looking for extra marks?*
>
> If LINEs and SINEs insert into functional genes they cause mutations. Cases of haemophilia A and B, X-linked severe combined immunodeficiency, and muscular dystrophy have all been linked to insertion of an *Alu* sequence (a SINE) in the relevant gene.

Class II transposons

- Class II transposons encode the enzyme **transposase** which catalyses excision of a segment of DNA and its subsequent insertion into a new position in the genome.
- During excision transposases recognize and bind to the inverted repeats at each end of a class II transposon.
- For insertion, some transposases require a specific sequence as their target site; others can insert the transposon anywhere in the genome.
- A staggered cut is made in the DNA at the target site, like the 'sticky ends' produced by some restriction enzymes.

  ➲ *See section 14.1 for details of restriction enzymes.*
- After the transposon is ligated into the host DNA, the gaps are filled in by a DNA polymerase and ligase. This creates the identical direct repeats at each end of the transposon (Figure 7.9).

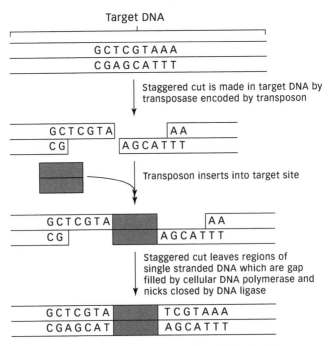

Figure 7.9 Generation of the direct repeats which flank class II transposons.

Example	Comment
Bacterial insertion sequences (IS)	Typical class II transposon: terminal inverted repeats either side of DNA coding for one or two transposases
Bacterial composite transposons (*Tn*)	Two IS transposons flanking coding DNA, e.g. *Tn10* – tetracycline resistance gene flanked by two IS10 sequences
P elements of *Drosophila*	Used as a vector for production of transgenic flies. An early embryo is injected with a P element containing the desired gene. As P elements lack functional transposase, an additional plasmid with an integrase is also injected
Ac and *Dc* elements of maize	The first transposable elements ever identified; discovered by Barbara McClintock. Pigmentation of maize kernels is controlled by the presence of *Dc*. Insertion into the C gene abolishes colour. *Dc* is non-autonomous and requires a tranposase from an *Ac* element for transposition
Miniature inverted-repeat transposable elements	Found in range of eukaryotic genomes, e.g. humans, *Xenopus*, apples, rice, *Caenorhabditis elegans*. Consist of almost identical sequences of ~400 bps flanked by inverted repeats of ~15 bps

Table 7.3 Class II transposons

- Class II transposons often lose their gene for transposase. This is generally the result of earlier incomplete excision of a transposon.
- These non-autonomous class II transposons use a transposase produced by another fully competent class II transposon to recognize their inverted repeats and move them to a new genomic location.
- Examples of class II transposons are presented in Table 7.3.

Looking for extra marks?

A group of bacterial transposons, conjugative transposons (CT_n), excise from DNA and form a circular intermediate, which transfers by non-pilus-mediated conjugation to another bacterial cell. They carry and spread antibiotic resistance genes. Positively, they also spread nitrogen fixation genes.

Function of transposable elements

- The large numbers of transposable elements in both prokaryotes and eukaryotes prompts the question as to whether they have any positive function.
- Table 7.4 considers detrimental and possible beneficial consequences of the replication and spread of transposable elements.
- Cells can inhibit most transposable element activity.
- Epigenetic silencing by DNA methylation, chromatin remodelling, and siRNA silencing mechanisms all prevent transposition.
- The ability of transposable elements to increase genetic variability (Table 7.4) considered together with a cell's ability to silence their activity, suggests a balance between deleterious and beneficial effects such that transposable elements have an important evolutionary role.

Detrimental effects	Possible benefits
Transposable elements are mutagens. Insertion into exons, introns, promoters, or enhancer DNA can destroy or alter a gene's activity	Transposable elements may help drive evolution by generating new variation, e.g. imperfect excision by a class II transposon can result in **exon shuffling** when the juxtaposition of two previously unrelated exons creates a novel gene product
Faulty repair of the gap left when a class II transposon is cut from a site can lead to mutation	Alteration of a gene's regulatory region by insertion of a transposable element may produce a new advantageous phenotype
Transposable elements increase the bulk of DNA to be replicated; constituting a metabolic burden to a cell	
The presence of tandem repeats of transposable elements disrupts precise pairing during meiosis. Unequal crossovers can lead to gene duplications	

Table 7.4 What good are transposable elements?

Looking for extra marks?

Integrons are another type of mobile genetic element. They capture genes, generally antibiotic resistance genes, by site specific recombination. They differ from transposons in lacking direct and indirect repeats at their ends.

Check your understanding

7.4 Describe, with examples, the different types of repetitive DNA.

7.5 Consider the bacterial transposon Tn9. The inverted repeats at each end of a Tn9 are the same for every Tn9 transposon, yet the direct repeats flanking an inserted Tn9 are different for each inserted Tn9. Explain.

7.6 What are the differences in the way class I and class II transposons move around a genome?

7.8 DNA REPLICATION

- DNA replication is a key process for all living organisms.
- The genetic information is stored in the sequence of DNA nucleotides.
- Before a cell divides this information is copied to produce new DNA molecules to pass to the new cells.

Semi-conservative nature of DNA replication

- During DNA replication, the two strands of a DNA molecule separate. Each strand functions as a template for synthesis of a new complementary strand.
- Each new DNA molecule contains an original and a newly synthesized strand, i.e. the original DNA molecule is half (semi-) conserved during replication (Box 7.1).

Box 7.1 Meselson and Stahl's demonstration that DNA replication is semi-conservative

- Meselson and Stahl used two isotopes of nitrogen (normal ^{14}N and the rare heavy ^{15}N) to distinguish between original and newly synthesized DNA.
- Initially, they grew *Escherichia coli* in media with ^{15}N as the sole nitrogen source so all DNA bases contained the heavier form of nitrogen.
- At time 0 they transferred a sample of bacteria from ^{15}N to ^{14}N culture and sampled over succeeding generations.
- Extracted DNA was subjected to caesium chloride density gradient, which separates DNA according to its density.
- The positions of the replicated DNA molecules in the density gradient confirmed the semi-conservative nature of DNA replication (see following diagram).

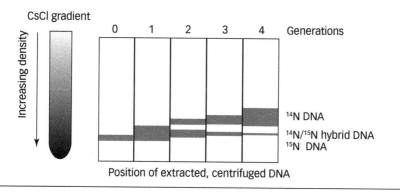

- The specificity of base pairing (adenine with thymine and guanine with cytosine) ensures only one sequence of nucleotides can be produced opposite a template strand.
- Thus, the two new DNA molecules are identical in their nucleotide sequence (information content) to the original.
- Matthew Meselson and Franklin Stahl convincingly demonstrated semi-conservative replication in a key experiment in 1958 (Box 7.1).

Details of the replication process

There are four main stages to DNA replication:

- initiation
- unwinding
- elongation
- termination.

DNA replication

(i) Initiation
- DNA replication begins at specific locations in the genome termed **origins**.
  - In a small prokaryotic genome there is a single origin.
  - Larger eukaryotic chromosomes contain multiple origins.
- Origin sequences are **AT-rich**, i.e. have proportionately more AT nucleotide pairs. Bases A and T bond by two hydrogen bonds, compared with the three bonds holding together C and G. Less energy is required to part DNA strands bonded by AT pairing.
- In *Escherichia coli* a single initiator protein, **dna A**, binds to the replication origin (**oriC**).
- A multi-unit **origin recognition complex** (ORC) recognizes and binds to origin sequences in eukaryotes.
- Binding of initiator proteins causes a short section of DNA to unwind, allowing the access of other proteins.

(ii) Unwinding
- **Helicases** are recruited to the origin sequences to unwind the DNA for replication.
- Helicases use the energy from ATP hydrolysis to break the hydrogen bonds between base pairs of opposite DNA strands.
- This action produces two single strands of DNA; each serves as a template for synthesis of a new complementary DNA strand.
- **Single-strand binding proteins** stabilize the separated strands. They:
  - protect the single-stranded DNA from nucleases
  - prevent the separated single DNA strands forming secondary structures, such as hairpins, which would inhibit the action of DNA polymerase.
- Unwinding and separation of the two DNA strands creates a **replication fork**, the Y-shaped structure at the site of DNA replication consisting of the two unwound DNA strands branching out from the double helix (Figure 7.11).
- As replication proceeds helicases continue to unwind the double helix and the replication fork moves along the DNA molecule in a 3′ to 5′ direction.
- This movement forces the DNA ahead of the fork to rotate. **Supercoils** build up in the DNA helix (see Figure 6.1). If unresolved these would eventually stop progress of the replication fork.
- **Topoisomerases** relieve these physical tensions by catalysing the transient breaking and re-joining of DNA strands.
- There are two classes of these enzymes:
  - type I topoisomerases cut one strand of the helix, enabling the cut strand to swivel around the uncut one and so remove the built-up coils
  - type II topoisomerases (or **DNA gyrase**) cut both strands at the same time, enabling one helix to pass through the other.
- Two replication forks develop from each origin of replication. The opened chromosomal area between two forks is termed a **replication bubble**.

(iii) Elongation

- **DNA polymerases** catalyse the synthesis of a new DNA strand.
- They catalyse the systematic addition of dNTPs complementary to the exposed template strands.
- A **phosphodiester bond** is formed when the 5′-phosphate group of the incoming nucleotide joins to the 3′-OH group of the preceding nucleotide on the growing strand.
- Bond formation involves nucleophilic attack by a lone pair of electrons of the 3′-OH of a newly incorporated nucleotide onto the alpha phosphate of an incoming nucleotide (Figure 7.10).
- The energy for phosphodiester bond formation comes from the hydrolysis of two of the three phosphates of the incoming dNTP. Pyrophosphate (PP_i) is released.
- No DNA polymerase can start DNA synthesis *de novo* on a bare template.
- All DNA polymerases require a nucleotide with a free 3′-OH group to which another nucleotide can be added, i.e. all DNA synthesis must be **primed**.
- During DNA replication a **primase** first synthesizes a short stretch of RNA, 10–12 ribonucleotides long.
- This **primer RNA** provides a 3′-OH group to which a DNA polymerase can add a DNA nucleotide. The priming RNA is subsequently removed.

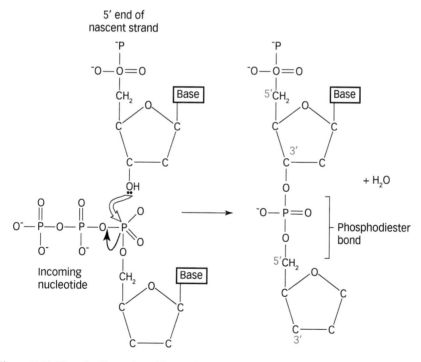

Figure 7.10 Phosphodiester bond formation.

DNA replication

- DNA replication proceeds differently on the two template strands. This is because of:
  - (i) The antiparallel orientation of the two DNA strands
  - (ii) 5′ to 3′ direction of DNA synthesis by DNA polymerases—they can only add a nucleotide to a 3′-OH group.
- On one template strand, the **leading strand**, polymerization is continuous in the 5′ to 3′ direction.
- On the other strand, the **lagging strand** polymerization is discontinuous (Figure 7.11A).
- Polymerization on the lagging strand proceeds from multiple RNA primers. It proceeds in a 5′ to 3′ direction, although it will appear as if it is happening in the opposite direction! (Study Figure 7.11.)
- The multiple sections of newly synthesized DNA on the lagging strand are referred to as **Okazaki fragments**.
- Eventually, RNase removes the RNA primers on both the leading and lagging strands.

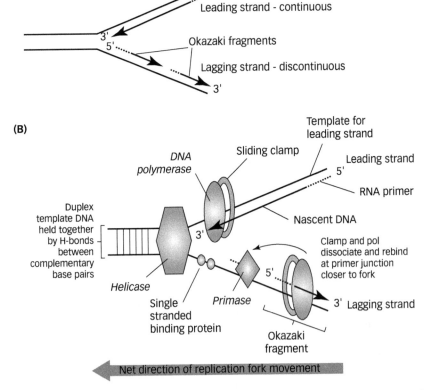

Figure 7.11 DNA replication at a replication fork. (A) The semi-discontinuous nature of DNA replication. (B) Details of enzymes and proteins involved in DNA replication.

- DNA polymerase catalyses the addition of complementary DNA nucleotides to fill the resulting gaps, leaving a single nick on the leading strand and multiple nicks on the lagging one.
- **DNA ligase** joins nucleotides at these nicks, completing the production of newly replicated DNA molecules.
- An alternative to replication at a replication fork is the **rolling circle** mechanism used by bacteriophages and plasmids to replicate their circular DNA genome (Table 7.5).

(iv) Termination

- In the circular prokaryotic chromosome there are specific termination sequences, e.g. **Ter sites** in a terminator region opposite oriC on the *E. coli* chromosome.
- Ter sites are 23 nucleotides in length and, when bound by **Tus protein**, stop replication.
- A Tus-coated Ter site stops movement of the helicase. This stalls the replication fork, which awaits the arrival of a convergent fork to complete replication.
- There are multiple Ter sites in the terminator region, e.g. seven sites in *E. coli*. One of these is used predominantly.
- Ter sites ensure ordered replication, i.e. that concurrent multiple rounds of the circular bacterial replication are not attempted.

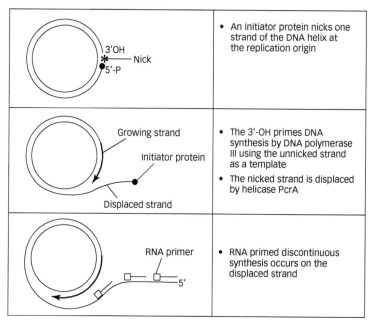

Table 7.5 The rolling circle mode of DNA replication

DNA replication

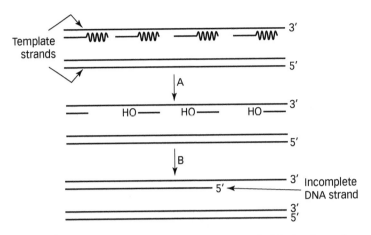

Figure 7.12 The telomere replication problem. (A) Removal of RNA primers from Okazaki fragments. (B) DNA polymerase synthesizes DNA to close gaps; final gap cannot be filled.

- There is no special termination mechanism on eukaryotic chromosomes. The multiple replication forks simply meet and merge at many points along a chromosome.
- DNA replication is, however, unable to continue to the very end of a linear eukaryotic chromosome on the lagging strand.
- There is a problem when the RNA primer is removed from the final Okazaki fragment: there is no free 3'-OH group to which to add dNTPs (Figure 7.12).
- As a consequence, small segments at each end of a chromosome are lost each time it is replicated.
- As the ends of eukaryotic chromosomes (known as **telomeres**) contain non-coding, repetitive DNA (e.g. human telomeres consist of ~2000 repeats of TTAGGG) this replication failure is not a problem.
- This steady shrinking of telomeres is believed to impose a definite lifespan on cells. After 40–60 cell divisions the telomeres become depleted and no further cell division is possible.
- A few cell types (germline and stem cells) need to maintain their telomeres. This is achieved with a **telomerase** (Figure 7.13).

Telomere replication

- A telomerase catalyses replication of a telomere.
- It is a **ribonucleoprotein** containing an inbuilt RNA sequence complementary to the telomere, in addition to its protein subunits.
- The telomerase RNA sequence is used as a template to extend the problematic telomeric template DNA strand (Figure 7.13).

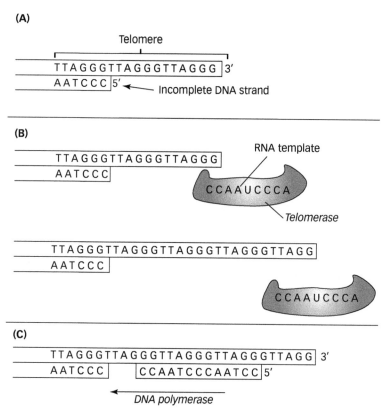

(A)

Telomere

TTAGGGTTAGGGTTAGGG |3'
AATCCC |5' ← Incomplete DNA strand

(B)

RNA template

TTAGGGTTAGGGTTAGGG
AATCCC

CCAAUCCCA

Telomerase

TTAGGGTTAGGGTTAGGGTTAGGGTTAGG
AATCCC

CCAAUCCCA

(C)

TTAGGGTTAGGGTTAGGGTTAGGGTTAGG |3'
AATCCC | CCAATCCCAATCC |5'

← DNA polymerase

Figure 7.13 Details of telomere replication. (A) Telomere with incomplete newly synthesized lagging strand. (B) Telomerase binds and extends the 3' of the template strand. (C) One final Okazaki fragment is synthesized.

- The truncated lagging strand is then completed by DNA polymerase alpha which has a DNA primase subunit to initiate DNA synthesis in a 5' to 3' direction.
- Thus, all the genetic information at the ends of linear chromosomes is copied.
- This is vital for germline and stem cells (both embryonic and somatic, such as in the bone marrow and intestine where continuous cell division is required).
- Activation of a telomerase-enabling telomeric replication also occurs in cancer cells and is believed to be a critical feature of their immortality.

Looking for extra marks?

The human premature ageing condition, Werner syndrome, is associated with shortened telomeres. Ageing symptoms manifest in children aged 6–10 years. It is an autosomal recessive condition, associated with mutations in the *WRN* gene on chromosome 8p12, encoding a DNA helicase.

Check your understanding

7.7 What are the roles of (i) RNA; (ii) Okazaki fragments; (iii) topoisomerase; and (iv) DNA ligase during DNA replication?

7.8 How are telomeres implicated in human ageing?

Fidelity of DNA replication

- DNA replication is extremely accurate, with an average error rate of 1 in 10^7 nucleotides. This high fidelity results from two functional aspects of the DNA polymerases:
  1. Accurate nucleotide pairing with selection errors only about 1 in 10^5 nucleotides.
  2. **Proofreading** by the polymerase. If an incorrect nucleotide is inserted, the mispaired nucleotides disrupt the active site of the polymerase. This triggers the 3′ to 5′ **exonuclease** activity of the DNA polymerase to remove the incorrect nucleotide. A new, correct one is inserted.
- A third process, **mismatch repair**, corrects errors remaining after replication. Incorrectly paired nucleotides deform the secondary structure of DNA, which is recognized by repair enzymes that excise and insert correct nucleotides.

  ⮕ *See section 11.4 (p. 223) for details of mismatch repair.*

DNA polymerases

- Cells possess a number of different polymerases: 5 in prokaryotes and 15 in eukaryotes.
- They are used variously in the different processes requiring DNA synthesis.
- In addition to DNA replication, synthesis of DNA occurs during repair of damaged DNA and recombination.
- Clamp proteins stabilize the interaction between DNA polymerases and the template strand (Figure 7.11).
- They enable polymerization rates of:
  - ~500 nucleotides/second by prokaryotic polymerases
  - ~50 nucleotides/second by eukaryotic polymerases.

Regulation of DNA replication

- Bacterial DNA replication is regulated through the combined action of:
  - hemi-methylation of the origin (oriC)

DNA polymerase	3' to 5' exonuclease	5' to 3' exonuclease	Functions
Escherichia coli pol I	√	√	• 5' to 3' exonuclease removes RNA primers of Okazaki fragments • Pol I then synthesizes replacement complementary DNA to complete replication
E. coli pol III	√	x	• Main prokaryotic replication enzyme • Synthesizes DNA in 3' to 5' direction from RNA primers on leading and lagging strands
Eukaryotic α	√	x	• Its primase synthesizes the RNA primer followed by the first 30–40 nucleotide stretch of an Okazaki fragment
Eukaryotic δ	√	√	• Main lagging strand enzyme • Synthesizes most of an Okazaki fragment
Eukaryotic ε	√	√	• Synthesizes the leading strand

Table 7.6 Key features of replication DNA polymerases

- ratio of ATP to adenosine diphosphate (ADP) in the cell
- levels of protein dnaA.
- These three factors influence binding of dnaA, the initiator of replication, to the origin.
  - DnaA binds less well to hemi-methylated DNA. It takes time for newly replicated DNA to be methylated.
  - ATP builds up in a cell during bacterial growth. It competes with ADP to bind to dnaA. ATP-bound dnaA binds to oriC.
  - Again, it takes time for dnaA levels to rise following DNA replication.
- In eukaryotes there are two key aspects to the control of replication:
  - DNA replication must only occur once during a cell cycle—during the S phase
  - replication must occur simultaneously from the thousands of replication origins in a eukaryotic genome.
- The replication licensing system, also called minichromosome maintenance complex, controls initiation of replication.
- During the G1 phase of the cell cycle, proteins are produced that bind to the origins of replication. This protein complex 'licenses' replication, i.e. when the correct molecular signals are received by these origin-bound proteins, replication is initiated.
- As the replication forks move away from the origins the licensing proteins are lost from the DNA. No new licensing proteins will be produced until division has occurred and the cell is once again in the G1 phase.
- A key component of the licensing complex is a DNA helicase that initiates unwinding of the DNA at the origin. Once replication has begun another protein, **geminin**, binds to the helicase preventing re-binding at the origin, and so initiation of another round of DNA replication.
- Geminin is degraded at the end of mitosis, enabling a new licensing complex to prime replication in the next cell cycle.

Check your understanding

7.9 DNA replication is semi-conservative. What does this mean?

7.10 You are synthesizing *E. coli* DNA *in vitro*. List the enzymes and proteins you will need and the various roles they will play in replicating this DNA.

7.11 Which of these DNA fragments will have a higher **melting temperature**? Why? (Circle one.)

(A) GCATTGACCGGAGGGACT   (B) GGATTTCAATTACTTAAT
    CGTAACTGGCCTCCCTGA   CCTAAAGTTAATGAATTA

 You'll find guidance on answering the questions posed in this chapter—plus additional multiple-choice questions—in the Online Resource Centre accompanying this revision guide. Go to www. oxfordtextbooks.co.uk/orc/thrive/ or scan this image:

8 From Genotype to Phenotype I: RNA and Transcription

Genes are expressed by being transcribed into RNA. Messenger RNA transcripts are translated into a polypeptide.

Key concepts

- RNA is an intermediary in the flow of information from genotype to phenotype.
- In living organisms genetic information passes from DNA to RNA.
- Different types of RNA perform different roles in the information transfer from genotype to phenotype.
- Transcription is the process of RNA synthesis from a DNA template.
- It is catalysed by RNA polymerases.
- Transcription involves three distinct phases: initiation, elongation, and termination.
- In eukaryotes there is extensive post-transcriptional processing of the primary RNA transcript.

8.1 INFORMATION TRANSFER

- Francis Crick (1958) stated that there is a one-way flow of genetic information from DNA to protein via the intermediary of RNA (Figure 8.1).

Structure of RNA

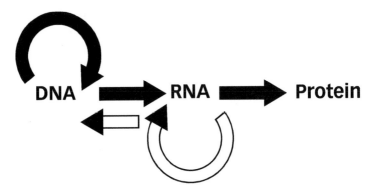

Figure 8.1 Information transfer in living organisms. Filled-in arrows represent DNA replication, transcription, and translation. Open arrows represent reverse transcription and RNA replication.

- The expression of different proteins directly or indirectly results in different phenotypes.
- Crick called this information flow from DNA to protein the **central dogma** of molecular biology.
- Subsequently, reverse flow (reverse transcription) was discovered, i.e. retroviruses transfer information from RNA into DNA.
- Information is also transferred from one DNA molecule to another via DNA replication.
- Some RNA viruses replicate their RNA genome and so transfer information from RNA to RNA.

8.2 TYPES OF RNA

- The majority of genes are transcribed into mRNA, yet this only accounts for 3–5% of total RNA.
- Other RNA types perform a variety of structural, regulatory, and enzymatic functions (Table 8.1).

8.3 STRUCTURE OF RNA

All RNA has the same primary structure. Different types of RNA possess various secondary structures.

Primary structure

- RNA is a polymer of ribonucleotides linked by phosphodiester bonds (section 7.3 and Figure 8.2).

Type of RNA	Function
mRNA (messenger RNA)	• Carries coding instructions for proteins
rRNA (ribosomal RNA)	• Major constituent of ribosomes (sites of protein synthesis) • In prokaryotes there are 3 rRNAs: 5S, 16S, and 23S • In eukaryotes there are 4 rRNAs: 5S, 5.8S, 18S, and 28S
tRNA (transfer RNA)	• There are 49 types of tRNA • Each tRNA carries a different amino acid, as specified by its **anticodon** loop, and holds it in place on the ribosome during translation
snRNA (small nuclear RNA)	• Involved in splicing (removal) of introns from pre-mRNA
snoRNA (small nucleolar RNA)	• Processes and chemically modifies rRNA
miRNA (microRNA)	• Regulates gene expression by inhibiting translation of mRNA
siRNA (small interfering RNA)	• Regulates gene expression by degrading mRNA • Helps establish compact chromatin structures
piRNA (piwi RNA)	• Silences retrotransposons in germ line cells
Other non-coding RNA	• Function in a variety of cell processes, e.g. telomere synthesis, X chromosome inactivation, transport of proteins into endoplasmic reticulum

Table 8.1 Types of RNA

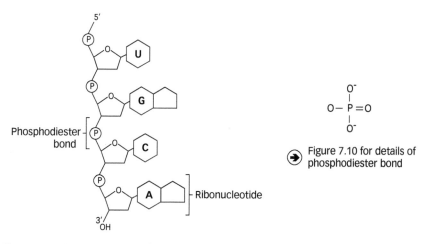

Figure 7.10 for details of phosphodiester bond

Figure 8.2 Primary structure of RNA: key features.

Secondary structure

- RNA readily forms secondary structures through complementary base pairing between single strands.
- Ribonucleotide sequence determines secondary structure.
- The most common secondary structure is a **hairpin loop**, also known as a **stem loop**.
- A hairpin loop forms when a **palindrome** is present, i.e. a ribonucleotide sequence that is repeated nearby in reverse complementary orientation.

Structure of RNA

- In a hairpin loop the palindromic sequences base pair to form a short double helix, termed the **stem**, that ends in an unpaired **loop** (Figure 8.3).
- Hydrogen bonding between the base pairs maintains RNA secondary structure: three hydrogen bonds between guanine and cytosine and two between adenine and uracil.
- Guanine–uracil pairings, involving two hydrogen bonds, are also found in RNA secondary structures.
- The stability of an RNA hairpin loop is determined by:
  - its length
  - the number of mismatches or bulges it contains
  - the base composition of the paired region—GC pairs with three hydrogen bonds are more stable
  - the size of the terminal unpaired loop, which is unstable and readily attacked by nucleases—optimal loop length is 4–8 bases.
- Three hairpin loops are a key feature of tRNA secondary structure (Box 8.1).

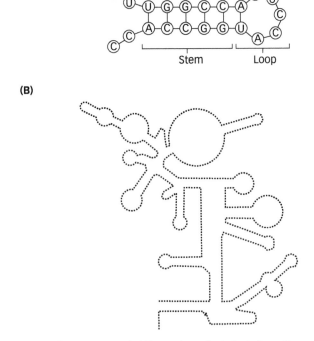

(A) RNA sequence: CUUGGCCAUUCCAUGGCCACC

Stem   Loop

(B)

Figure 8.3 RNA secondary structure. (A) Formation of a hairpin loop (bases in bold denote the palindrome). (B) Diagrammatic representation of the complex secondary structure of the RNA component of *Staphylococcus aureus*, RNase P.

Box 8.1 tRNA secondary structure

- The size and secondary structure of tRNA is similar in prokaryotes and eukaryotes.
- A mature tRNA molecule is composed of 73–93 nucleotides.
- The cloverleaf tRNA structure is produced by four base-paired **arms**, three of which end in unpaired loops (Figure 8.4).
  - (i) **Acceptor arm** formed from base pairing at the two ends of the single strand. A **CCA** triplet of unpaired nucleotides remains on the 3′ strand. This is the amino acid attachment site.
  - (ii) **TΨC arm.** Ψ represents pseudouridine (modified uracil).
  - (iii) **Anticodon arm** is vertically opposite the acceptor arm. It has a base-paired stem with a loop of seven unpaired nucleotides. The middle three nucleotides form the anticodon, which defines the amino acid carried by the tRNA and pairs with the mRNA codon.
  - (iv) **DHU arm**, or D loop, which generally contains dihydrouridine.
- There may be an additional **variable loop**.
- Stacking of the stem helices within the 'cloverleaf' produces a tertiary, L-shaped molecule.

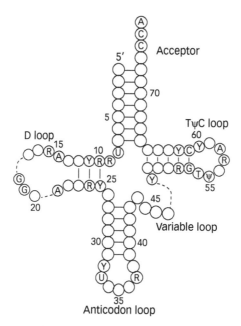

Figure 8.4 Two-dimensional cloverleaf structure of tRNA (T, Ψ, and D represent modified uracil to thymine T, pseudouridine Ψ, or dihydrouridine, R Y). Numbers refer to nucleotide positions.

- Hairpin loops are also important in:
  - ribozymes (catalytic RNA molecules)
  - prokaryotic rho-independent transcription termination (p. 160)
  - **attenuation** (p. 195).
- Ribozymes often form complex secondary structures with multiple hairpin loops, e.g. the RNA component of ribonuclease P (RNase P), that cleaves the 5′ end of precursor tRNA molecules (Figure 8.3B).

8.4 RNA POLYMERASES

- **RNA polymerases** catalyse RNA synthesis from a DNA template.
- RNA polymerases add a ribonucleotide to the 3′-OH group of an adjacent ribonucleotide.
- RNA synthesis is in the 5′→3′ direction, as during DNA synthesis.
- All types of RNA are synthesized by a single RNA polymerase in prokaryotes.
- Three main types of RNA polymerases synthesize eukaryotic RNA (Table 8.2).
- Two additional polymerases—RNA polymerase IV and V—have been found in plant cells. They synthesize siRNAs and RNA involved in determining chromatin structure respectively.
- RNA polymerases require various accessory proteins that bind and dissociate at different stages during transcription (sections 8.4 and 8.5).
- During transcription one DNA strand serves as the **template** for assembly of complementary ribonucleotides.
- The same base pairing rules apply as in DNA replication (C and G, T and A), with the additional pairing of uracil with adenine.
- The DNA strand used as the template strand during transcription is termed the **antisense strand**.
- The DNA strand not used as a template has the same nucleotide sequence as the RNA strand and is termed the **sense strand**.
- Different DNA strands function as the template strand for transcription of different genes.

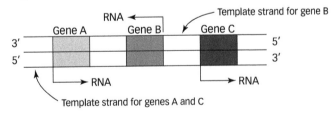

	RNA product
RNA polymerase I	18S, 5.8S, and 28S rRNA
RNA polymerase II	pre-mRNA, and most snRNAs and miRNAs
RNA polymerase III	tRNA, 5S rRNA, and other small RNAs

Table 8.2 Eukaryotic RNA polymerases

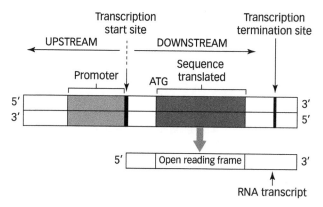

Figure 8.5 Key features of a transcription unit.

- Prior to initiating transcription RNA polymerases bind to **promoters** (Figure 8.5).
- A promoter is a specific DNA sequence upstream (away from the 5′ end) of a gene which signals to the RNA polymerase:
  - the direction of transcription
  - which of the two DNA strands to use as the template
  - the initiation site for transcription.
- RNA polymerase binding sequences within promoters are AT-rich.
- The A/T base pair is less stable than G/C, thus facilitating DNA **denaturation**, which enables access of RNA polymerase and accessory proteins to the template strand and so initiation of transcription.
- The synthesis of an RNA molecule following binding of the RNA polymerase at a promoter falls into three stages:
  1. Initiation—the RNA polymerase begins synthesis at an initiation site
  2. Elongation—RNA polymerase continues RNA synthesis, unwinding the DNA and adding new ribonucleotides in a 5′→3′ direction
  3. Termination—special terminator sequences signal RNA polymerase to cease synthesis and separate from the DNA template.
- The key features of initiation, elongation, and termination are similar in prokaryotes and eukaryotes, but eukaryotic transcription involves many more accessory factors.
- Transcription and translation are coupled in prokaryotes, i.e. ribosomes translate mRNA as it is being produced during transcription.
- In eukaryotes the two processes are compartmentalized. Transcription occurs in the nucleus, from which mRNA transcripts are exported into cytoplasm for translation following processing.

Checking your understanding

8.1 List eight functions of RNA.

8.2 What secondary structure is the sequence GCGCGUUAUUCGCGCAAUAA most likely to form, and why?

8.5 PROKARYOTIC TRANSCRIPTION

Bacterial RNA polymerases

- A bacterial RNA polymerase consists of:
  - a **core enzyme** of two α subunits, one β subunit, one β′ subunit, and one ω subunit
  - a sixth detachable **sigma** (σ) subunit, which binds the enzyme to the promoter to initiate transcription.
- This **sigma factor** is also referred to as sigma^{70} in reference to its molecular weight of 70 kDa.
- The core enzyme associated with the σ subunit produces the **holoenzyme**.
- The holoenzyme randomly binds and slides along DNA until the σ subunit encounters an available promoter. This prompts tight binding of the holoenzyme and initiation of transcription.

Bacterial promoters

- A bacterial promoter contains two critical **consensus sequences**, which are −10 and −35 nucleotides upstream from the gene.
- A consensus sequence is a stretch of common nucleotides found at the same location in different species and which are associated with key functions.
- The promoter **−10 consensus sequence**, also called the **Pribnow box**, is **TATAAT.** The recognition of this site is essential for the start of transcription in prokaryotes.
- The **−35 consensus sequence** is **TTGACA.** Its presence allows a high transcription rate.

- The sigma factor recognizes and binds to the −10 and −35 consensus sequences.
- This binding positions the RNA polymerase over a wider stretch of nucleotides— from −50 to +20 of the gene to be transcribed.
- Once tightly bound, the RNA polymerase initiates local unwinding of the DNA helix and alignment of the first ribonucleotide at the start site (+1) for transcription.
- The position of the −10 consensus sequence is the critical factor in determining where transcription begins, i.e. 10 nucleotides downstream.
- Experiments that move the consensus sequence correspondingly move the start point for transcription.
- This contrasts with eukaryotes, in which there is a specific transcription initiation sequence.

Looking for extra marks?

An alternative sigma factor is used when bacteria are heat shocked. An increase in temperature stimulates production of sigma^{32}, which recognizes the different promoter (with different −10 and −35 consensus sequences) of 'heat shock genes'.

Details of transcription

- To begin transcription the RNA polymerase pairs the base of a free ribonucleoside triphosphate with its complementary base on the DNA template strand at the start site. Unlike DNA synthesis no primer is required.
- A second ribonucleoside triphosphate is positioned, two phosphates are cleaved from the incoming ribonucleotide, and the first phosphodiester bond forms.
  ⊙ *See section 7.8 (p. 143) for details of phosphodiester bond formation.*
- Polymerization continues.
- Until the RNA polymerase clears the promoter, there is a tendency to produce short (2–6 nucleotides long) transcripts which are released from the DNA template. This is called **abortive initiation**.
- Once a growing RNA chain is 9–12 ribonucleotides long, the RNA polymerase has cleared the promoter. The sigma factor dissociates.
- Elongation is catalysed by the core enzyme in a **transcription bubble** of ~17 unwound nucleotide pairs.
- As the RNA polymerase moves down the DNA template, positive supercoils are generated ahead of the bubble and negative supercoiling behind. The stress of these supercoils is relieved by topoisomerases.
  ⊙ *See section 6.1 (p. 102) for details of topoisomerase relief of supercoils.*
- Elongation continues until the RNA polymerase transcribes a termination sequence, which signals the end of transcription.
- Termination sequences are encoded by the DNA but function by causing the formation of secondary structure within the RNA being transcribed. This destabilizes the RNA polymerase, causing its release.

Eukaryotic transcription

- Prokaryotes use two different strategies for transcription termination:
  - rho-dependent
  - rho-independent.

(i) Rho-dependent termination

- 'Rho' is an RNA-dependent ATPase which destabilizes the interaction between the template and the mRNA, promoting the release of the newly synthesized mRNA from the elongation complex.
- Rho binds at a cytosine-rich site on the nascent RNA strand, wraps around the RNA, and moves towards the RNA polymerase.
- When a stem loop structure within the terminator region pauses the polymerase, this is the signal for the **rho factor's** helicase activity to unwind the RNA/DNA complex. The nascent RNA is released and RNA polymerase dissociates, terminating transcription.

(ii) Rho-independent transcription termination

- The DNA terminator sequence is a GC-rich palindrome followed by a run of adenine nucleotides.
- The palindromic sequence causes formation of a GC hairpin loop in the RNA transcript. The resulting mechanical stress breaks the rU–dA bonds, pulling the transcript out of the RNA polymerase and terminating transcription (Figure 8.6).

> ### Check your understanding
>
> 8.3 What is the function of sigma factor and rho protein in prokaryotic transcription?

8.6 EUKARYOTIC TRANSCRIPTION

- Compared with prokaryotes, initiation of eukaryotic transcription is more complex.
- Eukaryotic RNA polymerases do not directly recognize the core promoter sequences. Instead, a collection of transcription factors mediate the binding of RNA polymerase and the initiation of transcription.
- Different RNA polymerases synthesize different RNA molecules (Table 8.2).
- Eukaryotic transcription initiation often involves regulatory sequences several kilobases away from the promoter.
- DNA upstream of a gene bends back on itself so regulatory sequences can access the RNA polymerase transcriptional complex.
- Eukaryotic transcription must deal with the accessing of DNA packed into nucleosomes and the 30 nm chromatin fibre—features absent from prokaryotic chromosomes.

The following section focuses on transcription by RNA polymerase II.

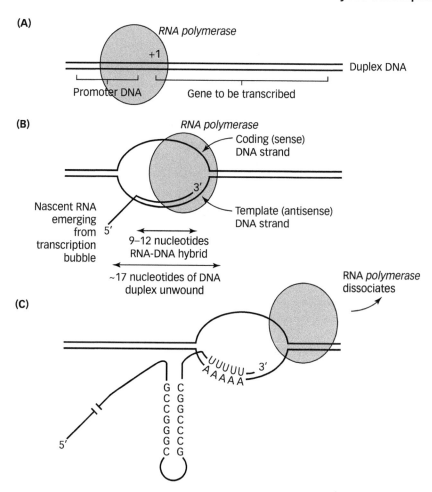

Figure 8.6 Transcription in prokaryotes. (A) Initiation; (B) elongation; (C) rho-independent termination.

Promoters

- There are two critical consensus sequences in two separate promoters.
  - The **core promoter** immediately upstream of the gene has the **TATA** box consensus sequence located around −30 nucleotides. The core promoter has additional consensus sequences (Figure 8.7).
  - The **regulatory promoter**, upstream of the core promoter, has the **CCAAT** box consensus sequence situated around −80 nucleotides from the start of the gene.
- Transcription factors and the RNA polymerase assemble into a basal transcription apparatus which binds to the core promoter and establishes a basal level of transcription.
- Transcriptional activator proteins interact with the regulatory promoter and **enhancer sequences**, often several kilobases upstream from the promoter.

Eukaryotic transcription

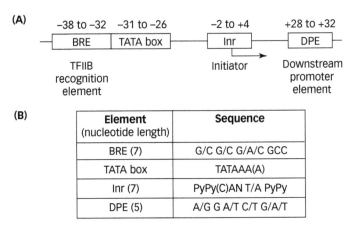

Figure 8.7 Core promoter elements: (A) organization of elements; (B) consensus sequences of elements.

- This interaction causes the DNA to bend back on itself so the activators can access both the enhancer sequences and the RNA polymerase transcriptional complex (Figure 8.8).
- Promoters of genes transcribed by RNA polymerases I and III possess different consensus sequences to those used by RNA polymerase II.
- The RNA polymerase III promoter is an **internal promoter** (i.e. downstream from the start of the gene) and so becomes part of the transcript.

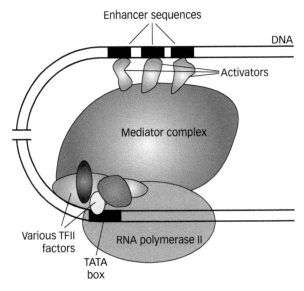

Figure 8.8 The role of the mediator complex and transcription factors in initiation of transcription.

Initiation

- The organization and control of transcription can seem very complex. Firstly, consider these basic requirements:
  - ○ the RNA polymerase must bind in the right place—at the start of a gene to be transcribed—and at the correct time in terms of the need for product developmentally/temporally
  - ○ the RNA polymerase must have full access to DNA—the DNA has to be freed from its association with histones in nucleosomes
  - ○ RNA polymerase must be positioned correctly on the DNA.
- During initiation of eukaryotic transcription six **transcription factors** (**TFIIs**) position RNA polymerase II on a **mediator complex** (a large multimeric protein) at the start of a protein coding sequence.
- The transcription factors are designated TFIIA, TFIIB, TFIID, TFIIE, TFIIF, and TFIIH, and are highly conserved in evolution.
- The functions of each factor are summarized in Table 8.3.
- The key stages in initiating transcription by RNA polymerase II are:
  - ○ **TFIID** recognizes and binds to the TATA box of the promoter by means of its **TBP** subunit; **TFIIA** stabilizes TFIID binding
  - ○ TFIID binding causes a large physical distortion of the DNA which signals other TFII factors to assemble
  - ○ **TFIIB** binds and positions **RNA polymerase II** helped by **TFIIF**
  - ○ **TFIIE** aids binding of **TFIIH**, which has two key functions:
    - ▪ adenosine triphosphate (ATP)-driven separation of DNA double helix at the transcription start point
    - ▪ phosphorylation of C-terminal domain of RNA polymerase II so that it is released from the transcription factors and can begin elongation.

Factor	Number of subunits	Role in transcription initiation
TFIIA	3	• Stabilizes TFIID
TFIIB	1	• Recognizes and binds to promoter BRE sequence (Figure 8.7) • Positions RNA polymerase II at the start site of transcription
TFIID TBP subunit TAF subunits	1 ~12	• Recognizes promoter TATA box • Assists transcription activation, e.g. TAFII-250 subunit has histone acetyltransferase activity, which relieves binding between DNA and histones in nucleosomes • Some eukaryotic promoters lack the TATA box; instead, TFIID recognizes downstream Inr and DPE sequences (Figure 8.6)
TFIIE	2	• Attracts and regulates TGIIH
TFIIF	3	• Stabilizes RNA polymerase interaction with TBP and TFIIB; helps attract TFIIE and TFIIH
TFIIH	9	• Unwinds DNA and phosphorylates RNA polymerase II

Table 8.3 RNA polymerase II transcription factors

Eukaryotic transcription

- The TFII factors/RNA polymerase II/mediator complex forms the *core* transcription pre-initiation complex (**PIC**), but additional factors are required to:
  - (i) regulate gene expression in terms of both the rate and timing of transcription
  - (ii) remodel chromatin so that the RNA polymerase has access to DNA.
- Transcriptional **activators** regulate the rate and pattern of transcription. They recognize and bind to enhancer sequences, often thousands of nucleotides upstream of the promoter (for more details see Chapter 10).
- They are brought into contact with the RNA polymerase complex via the mediator complex, a multi-subunit complex that functions as a transcriptional coordinator. Its large surface area enables many protein–protein interactions.
- The mediator complex also binds chromatin and histone modifying enzymes so that chromatin is remodelled and RNA polymerase can access DNA.
- The role of the mediator complex is summarized in Figure 8.8.

Elongation

- With the PIC assembled at the core promoter, RNA synthesis can begin.
- TFIIH uses its helicase activity to unwind DNA. Strand separation starts from about −10 bp (from the transcription initiation site).
- An ~17 bp transcription bubble is enclosed by the RNA polymerase, with ~8 ribonucleotides bound to the DNA template strand.
- RNA polymerase II uses ribonucleoside triphosphates (rNTPs) to synthesize an RNA transcript.
- rNTPs are positioned in a 5′ to 3′ direction, according to the base sequence of the DNA template strand. Phosphodiester bonds form between adjacent rNTPs.
- RNA polymerase II is a large, much-studied multi-subunit (10–12) enzyme.
- As with prokaryotic elongation:
  - supercoils develop in front and behind the transcription bubble, and are relieved by topisomerases
  - abortive initiation occurs.
- Once the transcript is about 30 nucleotides long, RNA polymerase II leaves the promoter and enters the elongation phase.
- Most TFII factors are left behind and quickly reinitiate another transcription event.

Termination

- RNA polymerase II continues to transcribe 0.5–2 kb downstream of the 3′ end of the last exon.
- There are no specific eukaryotic termination sequences.
- Instead, the 3′ end of mRNA is determined by post-transcriptional RNA processing (see section 8.6).

→ Check your understanding

8.4 What roles do promoters play during transcription?

8.5 List similarities and differences between transcription in prokaryotes and eukaryotes.

8.7 POST-TRANSCRIPTIONAL RNA PROCESSING

- Eukaryotic RNAs undergo major post-transcriptional processing before they are ready for use in translation:
  - all three main types of RNA (mRNA, rRNA, and tRNA) contain **introns** (intervening sequences) that need to be removed (**spliced**) before the RNAs are biologically active
  - additionally, mRNA is modified, or **capped**, at both the 3′ and 5′ ends.
- In contrast, bacterial RNAs are ready for use as soon as they are synthesized. Indeed, translation of mRNA often begins before transcription is completed.

Processing of mRNA transcripts

- The different stages of processing the pre-mRNA transcriptional product of a protein-coding gene ready for translation of its coding sequence are summarized as follows.
- When ready for translation by a ribosome, a fully processed eukaryotic mRNA molecule contains (in the following order):
  - 5′ methyl cap
  - **5′ untranslated region** (5′UTR)
  - the coding sequence (with all introns removed)
  - 3′ untranslated region (3′UTR)
  - polyA tail.
- Specific sequences in the 5′UTR and, particularly, the 3′UTR are bound by various proteins and small RNA molecules that regulate aspects of gene expression, e.g.:
  - mRNA localization within a cell
  - mRNA stability
  - translational efficiency, including inhibition.
  - → *See section 10.4 for details of regulation of gene expression.*
- Eukaryotic mRNA is **monocistronic**, i.e. it encodes a single polypeptide chain, in contrast with prokaryotic mRNA, which is generally **polycistronic**, encoding several polypeptides (Figure 8.9).
- In prokaryotes, a set of functionally related genes, an **operon**, is transcribed into a single mRNA molecule which contains several coding sequences (also referred to

Post-transcriptional RNA processing

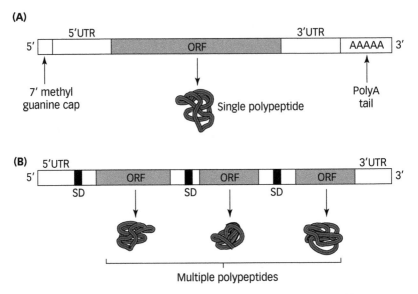

Figure 8.9 Structure of (A) eukaryotic mRNA and (B) prokaryotic mRNA (ORF: open reading frame; SD: Shine–Dalgarno sequence).

as open reading frames, ORFs). Each ORF has a start and a stop codon and is translated into a separate polypeptide.

- Between the ORFs and at each end of the mRNA molecule are UTRs. Again, they contain regulatory sequences.

- For example, before the start of each ORF is the **Shine–Dalgarno** ribosome binding sequence.

Processing of tRNA and rRNA transcripts

- The final tRNA molecule is cleaved from a larger precursor molecule. For example, *Escherichia coli* tRNA^tyr is 77 nucleotides long and is cleaved from a 126-nucleotide precursor.

- The ribozyme RNase P (Figure 8.3) universally processes precursor tRNA.

- Seven to 15% of tRNA nucleotides are modified. For example, guanine is methylated to form methylguanine and adenine deaminated to form inosine. The presence of modified bases enables specific interactions with amino acyl synthetases.
  ➜ *See section 9.4 (p. 177) for details of amino acyl synthetases.*

- Base modification also protects the tRNA molecule from degradation by **ribonucleases.**

- Internal base pairing produces the secondary cloverleaf structure which then folds into an L-shaped tertiary structure (Figure 8.4).

- Functional rRNA molecules are also cleaved from larger transcripts (see Figure 9.6).

8.8 DETAILS OF mRNA CAPPING AND SPLICING

mRNA capping

- The 5′ cap consists of an extra methylated guanine nucleotide.
- Fifty to 250 adenine ribonucleotides added to the 3′ end of the newly synthesized mRNA form the 3′ cap or **poly(A) tail**.
- Capping has various functions.
  - The 5′ cap:
    - signals the 5′ end of mRNA to subsequent RNA splicing complexes
    - binds a protein complex called cap binding complex (CBC), which aids in export of mRNA from the nucleus
    - enables ribosomes to recognize the beginning of the mRNA during initiation of translation.
  - The 3′ cap:
    - forms part of the ribosomal initiation complex for translation (discussed in section 9.5, p. 179)
    - increases mRNA stability by protecting it from exonucleases.

5′ capping

- When the nascent mRNA is about 25 nucleotides long its 5′ end is **capped** (NB, RNA synthesis is 5′ to 3′).
- A **capping enzyme complex** (**CEC**) is bound to the RNA polymerase II tail. As soon as the 5′ end of the transcript emerges the CEC is transferred to it.
- Details of the 5′ capping process are shown in Table 8.4.

$5'_{ppp}N_pN_pN$ - - - - - - 3′	Details of process
$\xrightarrow{}$ P$_i$ GTP $\xrightarrow{}$ PP$_i$	• An **RNA terminal phosphatase** removes one of the three terminal phosphate groups at the 5′ end of the nascent mRNA • **Guanylyl transferase** forms an unusual 5′ to 5′ phosphodiester bond between a guanine nucleotide and the 5′ terminal nucleotide
$5'G_{ppp}N_pN$ - - - - - 3′ — Methylation $5'H_3CG_{ppp}N_pN$ - - - 3′ CH_3CH_3	• **Methyl transferase** adds a methyl group to the 7′C of the new terminal guanine nucleotide, forming 7′-methylguanine • Generally, methyl transferase also adds a methyl group to the 2′C of the ribose of the second and third nucleotides

Table 8.4 5′ Capping of nascent eukaryotic mRNA

Details of mRNA capping and splicing

3′ Capping

- 3′ capping involves cleavage of the 3′ end of the nascent mRNA followed by the addition of ~250 adenine ribonucleotides.
- The triggers for cleavage are the following sequences near the 3′ end of the nascent mRNA:
  - a **polyadenylation signal sequence** (5′-AAUAAA-3′)
  - followed 11–30 nucleotides downstream by (**5′-CA-3′**) and a GU-rich sequence (Figure 8.10).
- A cleavage complex transfers from RNA polymerase II to the nascent RNA.
- Cleavage of the RNA occurs at (5′-CA-3′).
- Polyadenylate polymerase (PAP), a DNA-independent RNA polymerase, adds adenine ribonucleotides to the new 3′ end of the RNA molecule.
- As the poly(A) tails are synthesized, poly(A) binding protein (PABP) binds and protects the 3′ end from ribonuclease digestion.

mRNA splicing

- Ninety-nine per cent of pre-mRNA introns are removed using a **spliceosome** (an snRNA–protein complex).
- A few introns **self-splice**.

Using a spliceosome

- A spliceosome is a large complex of five snRNAs (U1, U2, U4, U5, and U6) and almost 300 proteins.
- A spliceosome:

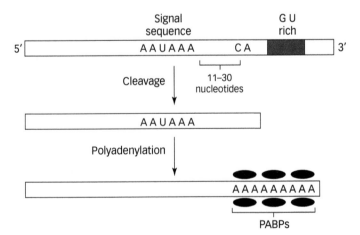

Figure 8.10 3′ RNA capping (PABP: to poly(A) binding proteins).

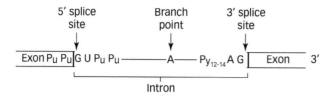

Figure 8.11 Critical pre-mRNA splice sequences. GU of the 5′ splice site is sandwiched between purine nucleotide. AG of the 3′ splice site is preceded by 12–14 pyrimidines.

- o detects the intron/exon boundary
- o cleaves RNA at appropriate points
- o joins adjacent exons together to produce mature mRNA.
- Splicing takes place in the nucleus before mature mRNA moves to the cytoplasm for protein synthesis.
- Critical consensus sequences are present at either end of an intron (Figure 8.11) termed the **5′ splice site** (or splice donor site) and the **3′ splice site** (or splice acceptor site).
- A crucial adenine nucleotide lies 18–40 nucleotides upstream from the 3′ splice site in the middle of a weak consensus sequence, generally rich in pyrimidines. This is the **branch point**.
- Splicing of an intron is a two-stage process (Figure 8.12).
  1. The pre-mRNA is cut at the 5′ splice donor site. This frees the 5′ end of the intron, which attaches to the branch point: guanine nucleoside from the 5′ end bonds with the free 2′ OH of the adenine nucleoside at the branch point. A **lariat** structure is formed.

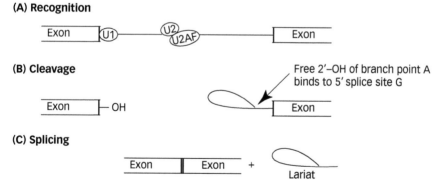

Figure 8.12 Splicing of a pre-mRNA intron: (A) the spliceosome's U1 snRNA recognizes and binds to 5′ donor site; U2 snRNA and binding protein U2AF recognize and bind to the branch point (U2 to nucleotide A and U2AF to the weak consensus adjacent polypyrimidine tract). (B) Cleavage and lariat formation; (C) formation of a new phosphodiester bond between exon boundaries.

Details of mRNA capping and splicing

   2. The 3′ splice acceptor site is cleaved and the 3′ end of the first exon is bonded to the 5′ end of the second. The intron is released as a lariat, linearized and degraded.

- Most mRNAs are produced from a single pre-mRNA from which the introns are removed and exons spliced together.
- In a few organisms (e.g. nematodes and trypanosomes) mRNAs are produced from two or more different pre-mRNAs. This is termed **trans-splicing**.
- **Alternative splicing**, the production of different mature mRNAs from the same pre-mRNA transcript depending upon which introns are spliced, occurs regularly.

   ⮕ *See section 10.4 (p. 203) for details of alternative splicing.*

Self-splicing

- During self-splicing, an intron forms a **ribozyme**: an RNA molecule that functions as a catalyst.
- Such introns assume a complex secondary structure with multiple hairpin loops (similar to RNase P in Figure 8.3).
- This structure catalyses its own excision from the RNA transcript.

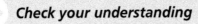

Looking for extra marks?

The human dystrophin primary transcript is unusually large, with 79 exons to splice together. A 2.4 Mb pre-mRNA transcript becomes a 14 kb functional mRNA.

   About 15% of single base substitutions that cause human inherited diseases affect mRNA splicing, resulting in abnormal processing of the primary transcript.

   For example, splice mutations are a common cause of beta-thalassaemia.

 Check your understanding

8.6 How are mRNA transcripts prepared for translation?

9 From Genotype to Phenotype II: RNA to Protein

The products of many genes are proteins or, more precisely, a single polypeptide, whose actions produce the traits encoded by their associated/encoding genes.

Key concepts

- Translation is the mechanism by which amino acids are assembled into proteins according to information encoded in mRNA.
- It takes place on ribosomes, which attach near the 5′ end of mRNA and move towards its 3′ end, positioning and linking amino acids according to the codons they encounter.
- Translation involves a number of RNA–RNA interactions between:
  - messenger RNA (mRNA) and ribosomal RNA (rRNA), which holds the mRNA for translation
  - mRNA codon and transfer RNA (tRNA) anticodon
  - tRNA and the rRNA of ribosomes.
- Translation in both prokaryotes and eukaryotes involves three main stages:
  - **initiation**—the components of the translation machinery are assembled at the start point near the 5′ end of mRNA

continued

- ○ **elongation**—ribosomes move towards the 3′ end of mRNA assembling and bonding amino acids
- ○ **termination**—synthesis halts at a stop codon and the newly formed polypeptide is released from ribosomes.
- Translation proceeds at an average rate of 20 amino acids/second in prokaryotes and 2 amino acids/second in eukaryotes.
- Protein synthesis consumes high levels of energy. Four high energy bonds (from guanosine triphosphate (GTP)) are hydrolysed per peptide bond formed.
- There is post-translational modification of polypeptides.

9.1 THE GENETIC CODE

- A gene's nucleotide sequence determines the amino acid sequence of a polypeptide.
- The sequence of three consecutive nucleotides encodes a particular amino acid.
- Each triplet of nucleotides is termed a **codon.**
- A code of 3 nucleotides per amino acid ensures all 20 amino acids are encoded separately. If one nucleotide encoded an amino acid the code would only have four variants; if two nucleotides, there would be 16 variants.
- There are 64 codons, yet only 20 amino acids. Thus, the code is **degenerate**, i.e. most amino acids are specified by more than one codon (Figure 9.1).
- Sixty-one codons specify the 20 amino acids. Three codons (UAA/UAG/UGA) are **stop codons** and signal termination of translation.
- The genetic code is **universal**, i.e. the same in all organisms. A rare exception is the stop codon UGA, which encodes tryptophan in the mitochondria of some species.
- The genetic code is continuous and **non-overlapping**, i.e. each nucleotide along a strand of mRNA is part of only one codon.
- For example, UGCACA is read as two consecutive, non-overlapping codons UGC ACA encoding cysteine and threonine and *never* as:

  UGC

    GCA

      CAC

        ACA encoding cysteine/alanine/histidine/threonine.
- There are three possible **reading frames**, or ways of translating an mRNA sequence. The correct reading frame is set by **initiation codon** AUG, encoding methionine.
- Consider mRNA sequence 5′ AUGCCAUUUGCA 3′.
- The correct, and only, reading frame is   AUG CCA UUU GCA

                           met pro phe ala

Ala	Arg	Asp	Asn	Cys	Glu	Gln	Gly	His	Ile	Leu	Lys	Met	Phe	Pro	Ser	Thr	Trp	Tyr	Val
	AGA									UUA					AGC				
	AGG									UUG					AGU				
GCA	CGA						GGA			CUA				CCA	UCA	ACA			GUA
GCC	CGC						GGC		AUA	CUC				CCC	UCC	ACC			GUC
GCG	CGG	GAC	AAC	UGC	GAA	CAA	GGG	CAC	AUC	CUG	AAA		UUC	CCG	UCG	ACG		UAC	GUG
GCU	CGU	GAU	AAU	UGU	GAG	CAG	GGU	CAU	AUU	CUU	AAG	AUG	UUU	CCU	UCU	ACU	UGG	UAU	GUU
A	R	D	N	C	E	Q	G	H	I	L	K	M	F	P	S	T	W	Y	V

Figure 9.1 The 20 amino acids (with their three- and one-letter abbreviations) and their mRNA codons.

Check your understanding

9.1 What are the key features of the genetic code?

9.2 What are the advantages and disadvantages of the universality of the genetic code?

9.2 KEY FEATURES OF PROTEINS

- Proteins are polymers consisting of a linear chain of amino acids linked by peptide bonds.
- All amino acids have a common chemical core and a variable R group (Figure 9.2).
- The 20 different amino acids that occur naturally in proteins are distinguished by their different R groups.
- The properties of the R group vary according to their chemical nature. The R group may be acidic or basic, polar or non-polar.
- The properties of the resulting protein relate to the R groups of their constituent amino acids. For example, histone proteins are rich in basic amino acids lysine and arginine, which enhance histones' interactions with negatively-charged DNA.
- During translation, amino acids are linked by a covalent **peptide bond** between the carboxyl group (–COOH) of one amino acid and the amino group (–NH$_2$) of the next amino acid (Figure 9.3).
- The linear chain of amino acids linked by peptide bonds is termed the protein's **primary structure**.

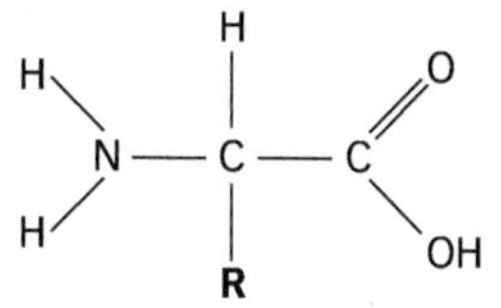

Figure 9.2 Generalized structure of an amino acid (shown in non-ionized form).

Key features of proteins

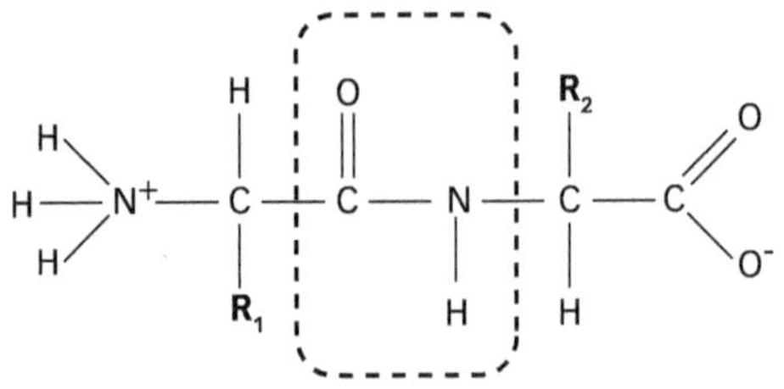

Figure 9.3 A dipeptide: two amino acids linked by a peptide bond (highlighted in dotted box).

- Interactions between the R groups of different amino acids shape the chain into secondary and tertiary conformations. Two or more polypeptide chains may interact to produce a quaternary structure (Figure 9.4).
- Two main types of **secondary structure** result from coiling and folding of the primary amino acid chain:
  - alpha helices
  - beta-pleated sheets.
- Hydrogen bonds stabilize the secondary structure.
- Proteins with a structural role tend to be functional in their secondary state, e.g. keratin.
- In many proteins the amino acid chain folds further (Figure 9.4).
- The resulting compact **tertiary structure** is stabilized by a variety of interactions between the R side chains:
  - disulphide bridges (involving amino acids cysteine and methionine)
  - ionic bonds
  - hydrogen bonds
  - van der Waals forces
  - hydrophobic interactions.
- Some proteins are functional in a **quaternary state**, i.e. as a multi-subunit protein.
- In addition, there may also be a non-protein **prosthetic group** associated with the polypeptide chains.
- For example, haemoglobin has haem prosthetic groups linked to each of its four globin subunits.

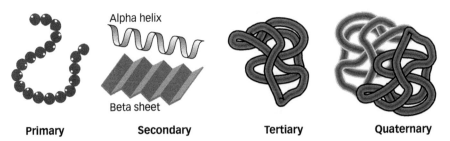

Figure 9.4 Summary of protein structure.

9.3 RIBOSOMES

Ribosomes are the sites of translation.

Structure of ribosomes

- A ribosome is composed of a small and a large subunit, which are separate unless a ribosome is actively synthesizing proteins.
- Each ribosomal subunit is a large complex of rRNA and many proteins.
- In a subunit, rRNA is folded into a highly compact core with the proteins largely located on the outer surfaces.
- Prokaryote and eukaryote ribosomes are similar in structure, although eukaryotic ribosomes are larger (Table 9.1).
- Mitochondrial and chloroplast ribosomes resemble prokaryotic ribosomes.
- A ribosome has three sites that are occupied, in turn, by a tRNA during elongation. These are the **aminoacyl** or **A site**; the **peptidyl** or **P site**; and the **exit** or **E site** (Figure 9.5; details in section 9.6).
- These sites are formed by rRNA, which also mediates the catalytic activity for peptide bond formation.
- The large subunit catalyses amino acid polymerization, while the small subunit facilitates the tRNA/mRNA interactions.

Cellular numbers and distribution

- There are typically 15,000 ribosomes in a single *Escherichia coli* cell, comprising ~25% of the dry cell mass.
- There can be millions of ribosomes in a single eukaryotic cell.

	Prokaryote ribosome	*Eukaryote ribosome*
Whole ribosome	70S* MW=2.5 MDa	80S MW=4.5 MDa
Large subunit	50S: 23S and 5S rRNA 31 proteins	60S; 28S, 5.8S and 5S rRNA ~49 proteins
Small subunit	30S: 16S rRNA 21 proteins	40S: 18S rRNA ~33 proteins

Table 9.1 Key features of ribosomal subunits.

*S is the sedimentation coefficient, measured in Svedberg units. The rate of sedimentation in a centrifuge is related both to the molecular weight and the three-dimensional shape of the complex. MW: molecular weight.

Ribosomes

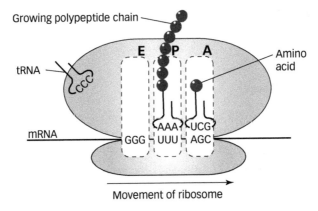

Figure 9.5 The three ribosomal sites (A, P, and E sites). In the A and E sites tRNA anticodons are bound to mRNA codons.

- Eukaryotic ribosomes may be found free in the cytoplasm or bound to the exterior of the endoplasmic reticulum (ER), which is termed 'rough endoplasmic reticulum'.
- ER-bound ribosomes are common in cells that package their newly synthesized protein for integration in the plasma membrane or secretion from the cell.
- Cytoplasmic ribosomes occur in greater numbers in cells that retain most of their manufactured proteins.
- Ribosomes may occur singly in the cytoplasm or in groups known as **polysomes**.
- Polysomes are formed from multiple ribosomes that are simultaneously translating one mRNA.
- In a polysome, ribosomes are spaced about 35 nucleotides apart on prokaryotic mRNA and 80 nucleotides in eukaryotes.

rRNA genes

- As huge amounts of rRNA are required in a cell there are many copies of the rRNA genes.
- Bacterial 16S, 23S, and 5S rRNA genes are organized into a multi-copy, single operon. *Escherichia coli* has seven rRNA operons.
- Eukaryotic 28S, 5.8S, and 18S rRNA is transcribed from a single tandemly repeated 45S unit. There are separate tandem arrays of the 5S rRNA gene.
- In humans, the 45S rRNA unit is repeated 300–400 times in five clusters on chromosomes 13, 14, 15, 21, and 22.
- The 45S unit has a characteristic organization. Internally-transcribed spacers (ITSs) separate the 28S, 5.8S, and 18S rRNA (Figure 9.6).
- During rRNA maturation, the external transcribed spacer and ITS regions are excised and degraded rapidly.

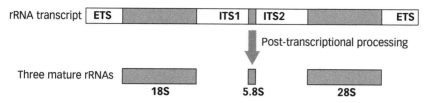

Figure 9.6 Organization and processing of 45S pre-rRNA. IST: internal transcribed spacer; ETS: external transcribed spacer.

> ## Looking for extra marks
>
> The DNA of ITS1 and ITS2 is frequently sequenced and analysed in molecular systematic and evolutionary studies. As ITS1 and ITS2 are spliced from the rRNA transcript there is weaker selective pressure to maintain sequence constancy, resulting in sufficient variable between different taxa to be evolutionary informative.

9.4 tRNA

tRNA carries amino acids to the ribosome during translation. The tRNA molecule folds into a characteristic 'cloverleaf' structure (see Box 8.1).

Binding of amino acids to tRNAs

- Binding of amino acids to tRNA is termed **tRNA charging** (Table 9.2).
- tRNAs are charged by enzymes called **aminoacyl-tRNA synthetases** in an energy-dependent two-step process (Table 9.2), which is summarized as:

  Amino acid + tRNA + ATP → aminoacyl-tRNA + AMP + PP_i

- There are 20 synthetases, i.e. one charging enzyme for each amino acid.
- **Isoaccepting tRNAs** (i.e. different tRNAs carrying the same amino acids) are charged by the same synthetase.
- The accuracy of charging of tRNA with the correct amino acid is crucial because, once charged, only the tRNA anticodon determines incorporation into the growing polypeptide chain.
- tRNA synthetases have **proofreading** capabilities, resulting in less than 1 in 10,000 amino acid charging errors.
- Each tRNA synthetase has an 'active-site pocket' for the correct amino acid.
- If an incorrect amino acid has bound to the enzyme, when the tRNA to be charged subsequently binds to the synthetase, an incorrectly bound amino acyl complex is formed. This aberrant complex is recognized and pushed into a second editing pocket and the incorrect amino acid removed.
- The three-dimensional shape of the tRNA is also a key factor in ensuring precise interaction between a tRNA and its synthetase. Specific tRNA nucleotides in the

tRNA

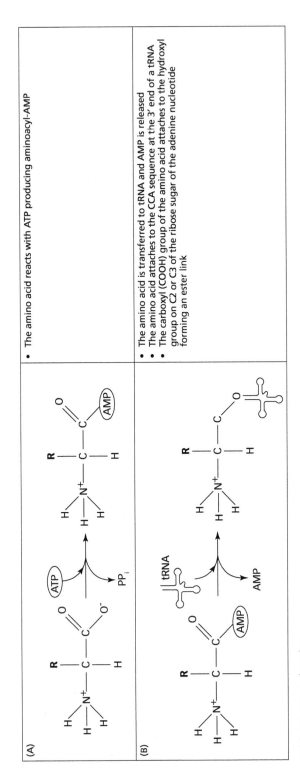

(A)
- The amino acid reacts with ATP producing aminoacyl-AMP

(B)
- The amino acid is transferred to tRNA and AMP is released
- The amino acid attaches to the CCA sequence at the 3′ end of a tRNA
- The carboxyl (COOH) group of the amino acid attaches to the hydroxyl group on C2 or C3 of the ribose sugar of the adenine nucleotide forming an ester link

Table 9.2 tRNA charging

tRNA anticodon: first position	mRNA codon: third position	Pairing
U	A or G	tRNA anticodon 3′ – X – Y – U – 5′ mRNA codon 5′ – Y – X – A/G – 3′
C	G	3′ – X – Y – C – 5′ 5′ – Y – X – G – 3′
G	U or C	3′ – X – Y – G – 5′ 5′ – Y – X – U/C – 3′
A	U	3′ – X – Y – A – 5′ 5′ – Y – X – U – 3′
I	U, C, or A	3′ – X – Y – I – 5′ 5′ – Y – X – U/C/A – 3′

Table 9.3 The wobble rule

D-loop, acceptor stem, and the anti-codon loop have been shown to be important in aiding different synthetases distinguish the correct tRNA among the multiple tRNAs.

Binding of tRNA to mRNA

- The tRNA anticodon recognizes and binds to the mRNA codon during translation.
- In all prokaryotes and eukaryotes there are fewer tRNA types than the 61 different mRNA codons encoding amino acids, e.g. 49 different tRNAs in humans.
- Some tRNAs recognize more than one mRNA codon. This is achieved by flexibility or **wobble** in the base pairing between nucleotide 3 of the mRNA codon and nucleotide 1 of the tRNA anticodon.
- Table 9.3 illustrates the wobble-rule:
  - inosine (modified tRNA adenine) pairs with multiple bases in the mRNA codon
  - guanine and uracil pair.

Check your understanding

9.3 How does a tRNA become charged with the appropriate amino acid?
9.4 An anticodon of a tRNA has the sequence 5′GCA3′. What amino acid does this tRNA carry?

9.5 DETAILS OF TRANSLATION: INITIATION

- In both prokaryotes and eukaryotes:
  - the small and large ribosomal subunits are initially kept separate
  - the small subunit is firstly positioned towards the 5′ end of the mRNA so that translation can begin at the AUG **start codon**

Details of translation: initiation

- various **initiation factors (IFs)** are involved:
  - three in prokaryotes
  - six in eukaryotes—eIFs 3 and 4 are large, multi-subunit complexes
- IFs recognize a special **initiator tRNA** (a methionine tRNA), used only in translation initiation
- there are acceptor and anticodon stem nucleotide differences between the methionine-encoding initiator tRNA and the methionine tRNA used in elongation.
- Table 9.4 gives details of initiation of prokaryotic translation.

Initiation in eukaryotes

- A **pre-initiation complex** (**PIC**) forms between the small ribosomal subunit, initiation factor eIF2, and the initiator tRNA carrying methionine.
- eIF2 directs binding of the initiator tRNA to the P site of the small ribosomal subunit.
- This PIC recognizes the methylguanine 5′ cap of mRNA and binds to the initiator **capping protein** (**eIF4E**) complexed with the mRNA 5′ cap.
- The poly(A) tail at the 3′ end of mRNA also plays a role in initiation (Figure 9.7).
- The polyA binding protein that covers the poly(A) tail attaches the tail to the initiation complex at the 5′ cap (Figure 9.7).
- Involvement of the poly(A) tail in initiation ensures degraded mRNA is not translated.
- Once the initiation complex is in place it moves along the mRNA until it locates the first AUG codon, which is signalled by its position in the **Kozak** sequence (5′ – ACCAUGG – 3′).
- As the initiation complex moves along the mRNA towards the start codon any secondary RNA structure is removed by eIF4A that functions as an ATP-dependent RNA helicase.
- When a Kozak sequence is encountered all eIFs dissociate and the large ribosomal subunit binds to produce a ribosome ready for protein synthesis.
- As with prokaryotic initiation:
  - the initiator tRNA is in the ribosomal P site
  - the A site is free for a second amino acid-charged tRNA to bind, leading to formation of the first peptide bond.
- In some genes the Kozak sequence is imperfect. This results in **leaky scanning** by the ribosomal initiation complex, i.e. it does not always stop at the first AUG, but continues until the second or third AUG codon.
- All these AUG codons are in the same reading frame.
- This is a eukaryotic mechanism to produce multiple gene products (**isoforms**) from a single mRNA (e.g. proteins with and without a signal sequence at their N-terminus) so that the same protein can be directed for use in different parts of the cell.

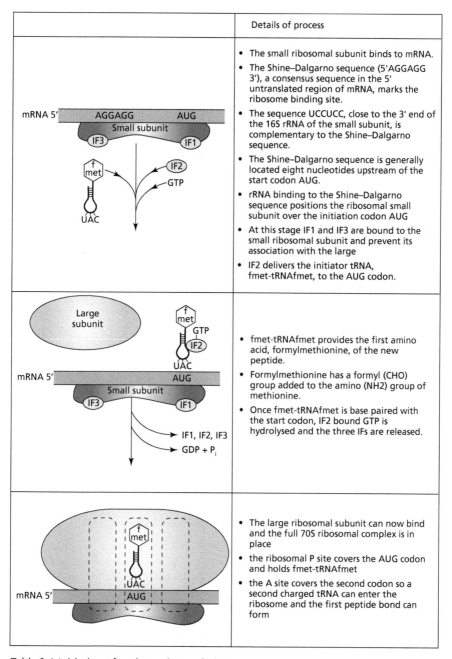

	Details of process
	• The small ribosomal subunit binds to mRNA. • The Shine–Dalgarno sequence (5'AGGAGG 3'), a consensus sequence in the 5' untranslated region of mRNA, marks the ribosome binding site. • The sequence UCCUCC, close to the 3' end of the 16S rRNA of the small subunit, is complementary to the Shine–Dalgarno sequence. • The Shine–Dalgarno sequence is generally located eight nucleotides upstream of the start codon AUG. • rRNA binding to the Shine–Dalgarno sequence positions the ribosomal small subunit over the initiation codon AUG • At this stage IF1 and IF3 are bound to the small ribosomal subunit and prevent its association with the large • IF2 delivers the initiator tRNA, fmet-tRNAfmet, to the AUG codon.
	• fmet-tRNAfmet provides the first amino acid, formylmethionine, of the new peptide. • Formylmethionine has a formyl (CHO) group added to the amino (NH2) group of methionine. • Once fmet-tRNAfmet is base paired with the start codon, IF2 bound GTP is hydrolysed and the three IFs are released.
	• The large ribosomal subunit can now bind and the full 70S ribosomal complex is in place • the ribosomal P site covers the AUG codon and holds fmet-tRNAfmet • the A site covers the second codon so a second charged tRNA can enter the ribosome and the first peptide bond can form

Table 9.4 Initiation of prokaryotic translation

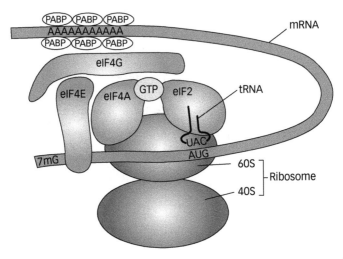

Figure 9.7 Initiation complex of eukaryotic translation: a summary diagram highlighting the role in initiation of the 3′ polyA cap.

 Check your understanding

9.5 How do (i) prokaryotic ribosomes and (ii) eukaryotic ribosomes recognize the 5′ end of the messenger RNA?

9.6 DETAILS OF TRANSLATION: ELONGATION

- In both prokaryotes and eukaryotes, amino acids are systematically added to the growing peptide chain in a cycle of three stages (Figure 9.8):
  1. tRNA binding
  2. Peptide bond formation
  3. Ribosome translocation one codon along the mRNA.
- Three elongation factors are involved:

Prokaryotes	Eukaryotes
EF-Tu	eEF1α
EF-Ts	eEF1βγ
EF-G	eEf-2

- A charged tRNA is already in the P site (Figure 9.8A). The P site tRNA holds the growing polypeptide chain.
- A tRNA carrying an amino acid enters the A site and its anticodon base-pairs with the mRNA codon.
- This tRNA is complexed with GTP and EF-Tu/eEF1 (Figure 9.8B).
- Once the A site tRNA is in place, GTP is hydrolysed and the resulting EF-Tu-GDP/eEF1-GDP is released (Figure 9.8C).

Details of translation: elongation

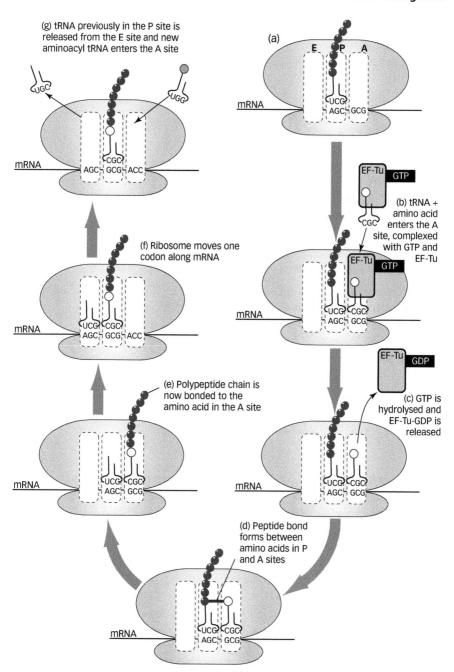

Figure 9.8 One elongation cycle during translation (details shown for prokaryotes; similar for eukaryotes).

- A peptide bond now forms between the amino acids that are attached to the tRNAs in the A and P sites.
- This bond formation is catalysed by the rRNA of the large subunit acting as a ribozyme. It releases the amino acid from the tRNA in the P site, which is now bonded to the tRNA in the A site.
- The ribosome moves one codon along the mRNA in the 5′ to 3′ direction. This requires EF-G/eEF 2 and hydrolysis of another GTP.
- As the tRNAs previously in the ribosomal P and A sites remain attached to mRNA this translocation moves the tRNAs to new sites in the ribosome (Figure 9.8E).
  - The tRNA previously in the P site now occupies the E site from which it is released into the cytoplasm to be recharged.
  - The A site tRNA now occupies the P site and a new amino acyl tRNA enters the A site.
- There are two opportunities for proofreading during elongation.
  - If there is an incorrect amino acid on a tRNA, EF-Tu/eEF1 does not bind properly to the aminoacyl tRNA. This incomplete binding impedes anticodon–codon pairing: the incorrect aminoacyl tRNA falls off.
  - Normally, once the anticodon is bound, EF-Tu/eEF1 causes two delays before the petidyl transferase can act:
    - it must hydrolyse GTP to GDP
    - it has to dissociate from the tRNA.
  - Both of these lags allow time for incorrectly bound tRNAs to fall off. Some of the correct tRNAs also fall off, but at a slower rate.

9.7 DETAILS OF TRANSLATION: TERMINATION

- Protein synthesis ceases when an mRNA stop codon occupies the A site.
- No tRNA enters as there is none with an anticodon matching a stop codon.
- Instead, **release factors** bind to the ribosome. These look like charged tRNAs (a phenomenon called molecular mimicry) and enter the A site, causing release of the polypeptide from the last tRNA and of this tRNA from the ribosome.
- There are three prokaryotic and one eukaryotic release factors (Table 9.5).

	Step I	*Step II*	*Step III*
Prokaryotes	RF1 binds UAA and UAG RF2 binds UAA and UGA	RF3 + GTP binds ribosome	- GTP hydrolysed - RF1, RF2, eRF1 released from A site - tRNA in P site moves to E site
Eukaryotes	eRF1 binds UAA, UAG, and UGA	eRF1 + GTP binds ribosome	

Table 9.5 Role of release factors (RFs) in termination

A number of antibiotics destroy bacteria by binding to their ribosomes, inhibiting different stages in translation, e.g.:

- tetracycline—blocks binding of aminoacyl tRNA to ribosome A site
- chloramphenicol—inhibits ribosomal peptidyl transferase activity
- streptomycin—elongation is inhibited
- erythromycin—binds to the E site of ribosomes.

9.8 TRANSLATION SUMMARY

- The mechanism of translating mRNA into protein is remarkably similar in prokaryotes and eukaryotes.
- Table 9.6 highlights similarities and the few significant differences.

Feature	Prokaryotes	Eukaryotes
Nature of ribosomes (see also Table 9.1)	• 30S and 50S subunits form functional 70S ribosome	• Larger • 40S and 60S subunits form functional 80S ribosome
rRNA	• Three types • 23S and 5S in large subunit • 16S in small subunit	• Four types • 28S, 5S, and 5.8S in large subunit • 18S in small subunit
Start codon	• AUG and occasionally GUG	AUG
Initiator tRNA	• Carries formylmethionine (fmet) • fmet is usually cleaved from mature protein	• Carries methionine (met) • Met is retained in mature protein
Ribosome binding to mRNA	• Shine–Dalgarno sequence recognized by 16SrRNA	• 5′ methylguanine cap complexes with small ribosomal subunit in pre-initiation complex
Finding of AUG start codon	• Small subunit binding over Shine–Dalgarno sequence includes AUG	• Small subunit moves 3′ from cap scanning for AUG in Kozak sequence
Initiation factors (IFs)	3 (IF1-3)	6 (eIF3 and 4 are multimeric)
Elongation factors	3 (EF-Tu, Ts, and G)	3 (eEF1α, eEF1βγ, and eEF2)
Release factors	3 (RF1-3)	1

Table 9.6 Key differences between prokaryotic and eukaryotic translation

Check your understanding

9.6 What roles are played by mRNA, tRNA, and rRNA during synthesis of a polypeptide by a ribosome?

9.7 Discuss the extent to which prokaryotic and eukaryotic ribosomes initiate translation in the same way.

9.9 POST-TRANSLATIONAL PROTEIN MODIFICATION

Following translation, polypeptides are processed to form functional proteins.

Forming a unique three-dimensional conformation

- As a polypeptide emerges from a ribosome it spontaneously adopts key secondary structures such as beta sheets and alpha helices.
- There are then three options for the newly synthesized polypeptide:
  - it folds unaided
  - it receives help from **chaperone** proteins, which both aid tertiary folding and prevent formation of mis-folded protein aggregates that are detrimental to cellular function
  - it is tagged by **ubiquitin** for degradation if large areas are mis-folded.
- Two groups of chaperone proteins (**heat shock protein** (Hsp) 60 and Hsp70) guide folding.
- Hsps were named following their discovery in response to a temperature rise, which greatly stimulates their production and hence a cell's ability to deal with high temperature-induced protein unfolding.
- **Hsp 70** routinely binds to exposed hydrophobic amino acids as a polypeptide leaves the ribosome, as hydrophobic patches readily induce misfolding.
- **Hsp 60** detects areas where folding has gone wrong and attempts to correct it. The chaperone forms a barrel-like complex, which holds the misshapen protein. It induces unfolding and enables a second attempt at correct folding.
- Abnormally exposed hydrophobic R groups indicate misfolding. Ubiquitin is added and the resulting polyubiquitin complex is moved to a proteosome and degraded. This is the fate of as much as one third of newly synthesized polypeptides.

Cleavage

- The initial N terminus formylmethionine of bacterial polypeptides and methionine of eukaryotes is removed.
- The first 13–36 amino acids at the N terminus of secretory or membrane-bound proteins often function to direct the protein to its cellular location. These amino acids form the **signal sequence** and are removed once the protein is in its correct cellular location.
- Many proteins are synthesized as inactive precursors that are activated under appropriate physiological conditions by limited proteolysis. This allows hormonal and enzymatic activity to be controlled with regard to time, space, and changes in environment.
- For example, folded proinsulin is produced following cleavage of the 24-amino acid signal peptide (see Figure 10.8). Proinsulin is further cleaved, yielding active insulin, which is composed of two peptide chains linked by disulphide bonds.

Chemical modification

- Following removal of the initiator methionine of a newly synthesized polypeptide, the new N terminus amino acid is often acetylated (an acetyl group, CH_3CO, added).
- Various amino acids along the polypeptide chain may be modified, e.g. phosphorylated or methylated.
- Carbohydrate chains may be added, forming glycoproteins.
- The tertiary or quaternary protein may be complexed with metal atoms.

Check your understanding

9.8 Following translation on a eukaryotic ribosome, discuss the various modifications that can occur to a newly synthesized polypeptide to produce a functional protein.

online resource centre You'll find guidance on answering the questions posed in this chapter—plus additional multiple-choice questions—in the Online Resource Centre accompanying this revision guide. Go to **www. oxfordtextbooks.co.uk/orc/thrive/** or scan this image:

10 Control of Gene Expression

Regulatory mechanisms ensure genes are only expressed when and where they are needed.

Key concepts

- The expression of most genes in prokaryotes and eukaryotes is regulated, i.e. genes are switched on and off according to a cell's needs. This prevents cells' resources being wasted.
- A low percentage of a cell's genes, those that encode basic cellular functions, are expressed continually.
- In prokaryotes, changes in gene expression tend to be in response to environmental signals.
- In eukaryotes, differential gene expression tends to be related to developmental stage.
- Gene expression can be regulated at various points between genotype and phenotype:
  - in prokaryotes, most control mechanisms regulate the transcription of genes
  - eukaryotes employ various translational, as well as transcriptional control mechanisms.
- Positive control mechanisms indicate that a gene is switched off until switched on by an activator molecule.

continued

- Genes under negative control are expressed unless switched off by a repressor.
- Eukaryotes tend to use positive control mechanisms more often, whereas prokaryotic genes are generally negatively controlled.
- Transcriptional regulatory proteins often bind to specific DNA sequences and so have characteristic DNA binding domains.

10.1 LEVELS OF GENE REGULATION

- Figure 10.1 summarizes the key stages in the production of a functional eukaryotic protein from its gene. Expression of a gene can be regulated at any of these stages.

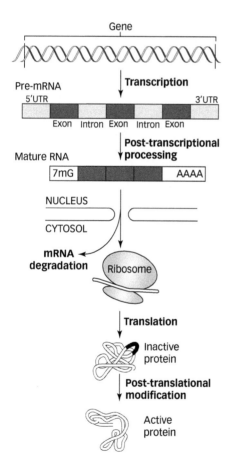

- Access of the transcriptional apparatus to a eukaryotic gene depends upon the relaxation of its chromatin structure, primarily achieved by chemical modification of nucleosome histones

- DNA methylation represses transcription.

- Various aspects of transcription are regulated:
  - access of an RNA polymerase to a promoter can be enhanced by activators or impeded by repressors
  - numerous transcription factors interact with RNA polymerases increasing or decreasing their affinity for a promoter

- Pre-mRNA processing is a major control step:
  - 5' capping and 3' tail addition determines RNA stability.
  - splicing of introns produces the functional mRNA. The variation of splice sites varies the nature of the gene product.

- The amount of protein product depends upon rates of mRNA degradation (high levels of degradation = lower levels of protein synthesis) as well as rates of mRNA synthesis.

- Both access to translation initiation sites and rates of translation are regulated.

- A final regulation point is often the pattern of chemical modification of the newly produced polypeptide, to produce a fully functional protein.

Figure 10.1 Multiple control points of eukaryotic gene expression.

10.2 DNA BINDING PROTEINS

- Transcriptional control involves regulatory proteins that bind to specific DNA sequences.
- DNA binding proteins are characterized by distinctive binding **domains**, which:
  - typically consist of 60–90 amino acids
  - are rich in amino acids arginine, asparagine, glutamine, glycine, and lysine
  - have a distinctive **motif** (Table 10.1) within their binding domain
  - forms hydrogen and other bonds with the bases of the DNA control sequence.

Motif	Description	Regulatory proteins
Helix-turn-helix	• Two alpha helices connected by a short irregular region, or 'turn' • The pattern of interactions between the amino acid R-groups of the C-terminal helix and bases within the major groove determine the sequence specificity of the DNA binding • This helix lies in the major groove of the DNA double helix	• Many bacterial regulators, e.g. catabolite activator protein (CAP)
Zinc finger	• A zinc ion stabilizes a loop of amino acids • One side of the loop is an alpha helix • The zinc ion is held in place by the R groups of two cysteine and two histidine residues • The alpha helix of the zinc finger lies in the major groove of the DNA double helix • Zinc finger motifs are often repeated in clusters	• Early growth response protein 1 (EGR-1)—a transcription regulatory factor
Leucine zipper	• An extension from each of two polypeptide monomers interact in parallel • Each extension is an alpha helix with a leucine every seven residues • Leucines interact to hold the dimer in a 'coiled coil' structure • The alpha helices diverge below the dimerization interface, allowing them to fit into the major groove of the DNA double helix	• Many eukaryotic transcription factors and developmental regulators
Helix-loop-helix	• A short and longer alpha helix connected by an irregular region, or 'loop' • Part of the motif dimerizes with a helix-loop-helix motif from another protein • Two sections of alpha helix, one from each monomer, bind to the major groove of the DNA double helix	• Various eukaryotic transcription factors
Homeodomain	• Three alpha helices, each connected by a short loop • The two N-terminal helices lie antiparallel • The longer C-terminal helix is perpendicular to this N-terminal pair • C-terminal helix interacts with the major groove of DNA	• Key regulators of development: initiate cascades of co-regulated genes, e.g. trigger expression of **homeobox** genes

Table 10.1 DNA binding motifs

- Regulatory DNA binding proteins frequently possess other domains that bind with other regulatory molecules.
- Their binding to DNA is often transient—part of dynamic cycles of binding and unbinding so that other proteins can compete and modulate gene expression.
- The DNA binding domain of regulatory proteins binds to the major groove of DNA.
- Within the major groove of DNA, each of the four DNA bases has a different pattern of hydrogen donors and acceptors, and hydrophobic groups on their exposed outer surface.
- Thus, a given DNA regulatory sequence presents a distinctive signal of attractive and repulsive forces that is recognized by a regulatory protein bearing a characteristic amino acid sequence within its DNA binding domain.
- A distinctive area of secondary structure within the DNA binding domains enables a regulatory protein to make intimate contact with DNA (Table 10.1 and Figure 10.2); this distinctive area is called a motif.
- Typically, there are 10–20 contact points between a regulatory sequence and its complementary regulatory protein involving hydrogen bonds, hydrophobic interaction, van der Waals forces, and ionic bonds: each weak, but collectively producing strong DNA binding.

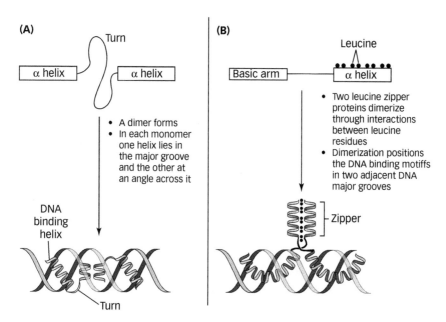

Figure 10.2 Two DNA binding motifs and their interaction with DNA. (A) Helix-turn-helix; (B) leucine zipper.

10.3 CONTROL OF GENE EXPRESSION IN PROKARYOTES

- Transcription is the main target for regulation of gene expression in prokaryotes.
- Most bacterial genes are functionally grouped into **operons** (Figure 10.3) whose transcription is regulated by the presence or absence of **repressors**.
- Expression of a few bacterial genes is regulated at the translational level. Ribosome binding to mRNA is controlled by:
  - the presence/absence of regulatory antisense RNA
  - the nature of secondary structure in the 5′ UTR of mRNA.
- The expression of operons is regulated in response to environmental levels of metabolites.
- Depending on the nature of the gene products, transcription of operons is either **induced** or **repressed**.
  - **Inducible operons** encode proteins that are only required occasionally, e.g. *Escherichia coli*'s *lac* operon. Transcription is inhibited until events 'induce' removal of a repressor (Figure 10.3).
  - Transcription occurs from a **repressible operon** until a repressor switches it off, e.g. *E. coli*'s *trp* (tryptophan) operon (Figure 10.4).
- Enzymes encoded by inducible operons generally function in catabolism, and their synthesis is induced by the presence of the substrate for digestion.
- Repressible operons usually encode enzymes needed for anabolic synthetic pathways and are inactivated by high levels of their end product.
- As repressors regulate both inducible and repressible operons (Table 10.2), albeit in different ways, repressor-mediated regulation is also termed **negative control**.
- These two types of transcriptional control are an **adaptive strategy** that ensures:
  - proteins in regularly high demand are present continuously (repressible operons)
  - proteins needed occasionally are only made when relevant (inducible operons).
- Each operon has a regulatory switch—the **operator**.
- The operator is a segment of DNA within the operon's promoter to which the repressors binds; binding of the repressor switches the operon off.
- When a repressor is bound to the operator it blocks progress of the RNA polymerase.
- A repressor is the product of a separate regulatory gene (Figure 10.3a).

(A)

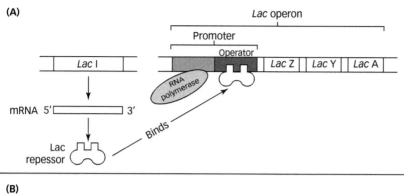

(B)

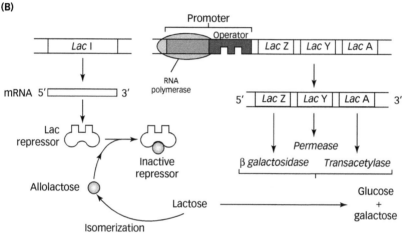

Figure 10.3 Control of *Lac* operon expression. (A) Absence of lactose: repressor is bound to operator and transcription inhibited. (B) Presence of lactose: allolactose binds to and inactivates repressor so that transcription occurs.

	Normal state of operon	Repressor
Inducible operons	Inactive	Active: inhibiting
Repressible operons	Active	Inactive: only inhibits when activated

Table 10.2 Negative control of operon expression

Expression control of an inducible and repressible operon is illustrated by the Lac and trp operons respectively.

Inducible operons: the Lac operon

- The products of an inducible operon are only required occasionally.
- Enzymes encoded by *lacZ*, *lacY*, and *lacA* enable bacteria to digest lactose on the rare occasions it is present in the environment.
- In normal conditions, when there is no lactose, a repressor is bound to the *Lac* operon operator, blocking access of the RNA polymerase to the promoter; consequently, there is no transcription.

Control of gene expression in prokaryotes

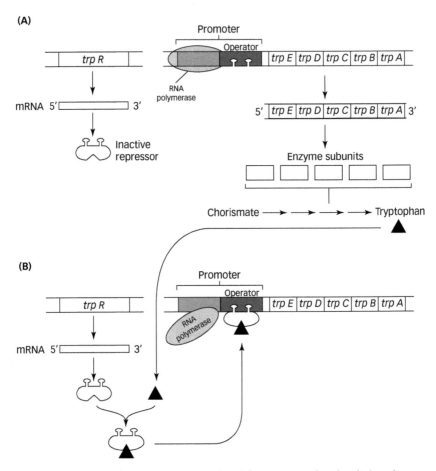

Figure 10.4 Control of *Trp* operon expression. (A) Low tryptophan levels: inactive repressor so transcription occurs. (B) High tryptophan levels: tryptophan binds to and activates the repressor, which binds to the operator and inhibits transcription.

- When lactose is present, some of it is converted to allolactose (an isomer of lactose), which binds to the repressor, inactivating it. Thus, allolactose is the **inducer** of *Lac* operon transcription.
- The repressor drops away from the operator; RNA polymerase can now access the promoter and transcription occurs (Figure 10.3).
- Removal of the repressor from the *Lac* operon establishes a basal level of transcription.
- The operon's rate of transcription is set by the level of cellular glucose, in a **positive control** mechanism:
  - when both lactose and glucose are present, *E. coli* preferentially metabolizes glucose
  - when the *Lac* operon is switched on, by allolactose binding to and inhibiting the *Lac* repressor, a low rate of transcription is established

- only if glucose levels are low is transcription of the *Lac* operon increased
- low glucose levels lead to increased cAMP levels; cAMP activates catabolite activator protein (CAP), which binds to the *Lac* promoter and increases the rate of transcription.

Repressible operons: the *trp* operon

- The products of a repressible operon are required continuously.
- Polypeptides encoded by the five genes *trpA–E* form the three enzymes used to synthesize tryptophan from chorismate.
- The amino acid tryptophan is a key constituent of proteins and so transcription of the *trp* operon occurs continuously.
- If tryptophan is present in the environment *E. coli* no longer needs to synthesize it.
  - Tryptophan binds with, and activates, a repressor.
  - The activated repressor binds to the operator and blocks access of the RNA polymerase to the promoter, thus inhibiting transcription (Figure 10.4).
- As with the inducible *Lac* operon, a second level of transcriptional control of the *trp* operon sets the rate of transcription in a **positive control** mechanism.
- Positive transcriptional control of the *trp* operon is by **attenuation**.
- This occurs when transcription is initiated, but terminates prematurely, before the structural genes are reached because of a secondary RNA structure that forms in the 5' untranslated region of the mRNA (Figure 10.5).
- A **leader sequence** is located between the operator and *trp E* gene in this 5' untranslated region.
- The leader sequence is 160 nucleotides in length; the resulting mRNA can form two alternative hairpin loops, one of which inhibits further transcription (Figure 10.5).

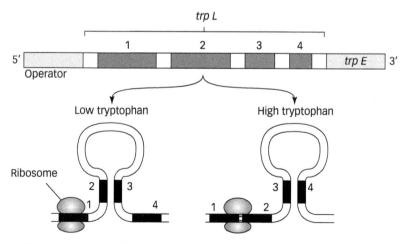

Figure 10.5 Attenuation of the *Trp* operon. At low tryptophan levels domains 2 and 3 base-pair and transcription rate is normal. At high tryptophan levels domains 3 and 4 base-pair and transcription is inhibited.

Control of gene expression in prokaryotes

To understand how attenuation controls the rate of transcription remember that translation of prokaryotic mRNA begins as it is being formed.

- The leader sequence (also termed *trp L*) contains four domains (Figure 10.5):
    - domain 1 encodes a 14-amino acid, non-functional leader peptide which contains two adjacent tryptophan codons
    - domain 3 base-pairs with domain 2 or 4, depending upon cellular tryptophan levels.
- When cellular tryptophan levels are high:
    - the ribosome rapidly translates domain 1 leader peptide
    - domain 2 becomes part of the ribosomal complex
    - domains 3 and 4 form a hairpin loop that terminates transcription.
- When cellular tryptophan levels are low:
    - ribosome stalls at the tryptophan codons in domain 1
    - this gives an opportunity for domains 2 and 3 to bind; the resulting hairpin loop favours continued transcription (Figure 10.5).

Looking for extra marks?

Bacteria also contain **regulons**, which are collections of genes scattered throughout the genome and controlled by the same factor or stimulus, e.g. heat. Regulons are also activated as part of quorum sensing, i.e. regulating bacterial growth in relation to local population density.

Riboswitches

- **Riboswitches** are regulatory sequences in an RNA molecule that assume different secondary structures in response to the binding of regulatory proteins.
- Riboswitch regulation is similar to attenuation control of the *trp* operon.
- The regulation of gene expression is achieved when the unfavourable secondary structure disrupts the ribosome binding site thus inhibiting translation.
- mRNAs that contain a riboswitch often regulate their own expression in a dosage-dependent mechanism. As the level of a protein product rises it binds the riboswitch, inhibiting further synthesis.
- For example, many bacterial species use a cobalamin riboswitch to regulate synthesis of vitamin B12 (cobalamin).
- As levels of cobalamin rise in a cell, molecules bind to riboswitch sequences in the 5'UTR of genes encoding enzymes used in its biosynthesis, preventing further translation of the genes.

Antisense RNA

- Antisense RNA (RNA with a sequence complementary to a certain target sequence) is a regulator of gene expression.

- It binds to the 5'UTR of mRNA, thus preventing ribosome attachment and translation.
- Antisense RNA is used to control expression of a variety of bacterial, phage, and plasmid genes.
- A bacterial example of antisense RNA regulation is control of expression of the outer membrane protein F (*omp F*) gene of *E. coli*.
- *OmpF* is a membrane transport protein that channels small, polar molecules, such as water, ions, and glucose through the cell membrane.
- If there is an increase in the osmolarity of the bacterial cell, synthesis of ompF decreases. This depresses transport of osmotically-active solutes, helping to rebalance cellular osmolarity.
- This decrease in *ompF* synthesis is achieved by antisense RNA:
  - regulator gene *micF* is activated and produces *micF* RNA
  - *micF* RNA binds to the 5'UTR region of *ompF* mRNA
  - this antisense RNA binding inhibits ribosome binding to *ompF* mRNA
  - fewer *ompF* channel proteins are synthesized, reducing transport of osmotically-active molecules and so helping to rebalance cellular osmolarity.
- In this example of antisense RNA regulation (and many others), the system enables protein dosage-dependent regulation, e.g. regulation of the ColE1 RNA primers for DNA replication of this plasmid.

Check your understanding

10.1 What is the difference between an inducible and a repressible operon?

10.4 CONTROL OF GENE EXPRESSION IN EUKARYOTES

- All cells of a eukaryote have the same genetic information, but express different sets of proteins.
- Individual cells express only a fraction of their genes.
- Differences in gene expression underlie the diversity of cell types in eukaryotes.
- Precise control ensures the appropriate genes are expressed in the relevant cells at the correct time during development.
- Eukaryotic gene expression is regulated at many different stages between DNA and functional protein (Figure 10.1).
- A few genes are **constitutive**, i.e. are not regulated, but are expressed continually.
- Constitutively expressed genes are called **housekeeping genes**. They encode proteins required for the maintenance of basic cellular function, e.g. the genes encoding the cytoskeletal protein actin and glycolytic enzyme glyceraldehyde 3-phosphate dehydrogenase (GAPDH).

Control of gene expression in eukaryotes

Regulating initiation of transcription

- As in prokaryotes, initiation of transcription is the key stage of control of eukaryotic gene expression. Transcriptional control is the least wasteful of a cell's resources.
- Unlike prokaryotes, with their operons, each eukaryotic gene is regulated separately.
- The regulation of transcription determines:
  - when transcription occurs
  - how much RNA is created.
- There are two main aspects to regulation of the initiation of transcription:
  - the binding of transcription activators to regulatory DNA sequences
  - changing chromatin structure so that activators have access to control sequences.

(i) Transcription activators

- Activators are modular proteins with different domains that:
  - recognize and bind to a DNA sequence (Table 10.1)
  - interact with one or more proteins of the transcriptional apparatus
  - interact with proteins bound to nearby regulatory sites
  - influence chromatin structure (directly or indirectly)
  - act as a sensor of conditions within the cell.
- Transcription activators bind to **enhancer sequences** within DNA. These regulatory sequences, also called **upstream activation sequences** (UAS), lie hundreds, and often thousands, of nucleotides upstream of a gene.
- A single gene frequently has multiple enhancer sequences, enabling it to respond to many activators.
- Enhancer-bound activators contact the mediator complex on the gene's promoter through looping segments of DNA (Figure 10.6).
- Transcription begins when activators, general transcription factors, RNA polymerase II, and other proteins are assembled with the mediator complex.
- RNA polymerase II and transcription factors assembled on the mediator complex at the core promoter (Figure 8.8) often establish a low, basal level of transcription.
- Transcriptional activators then set the rate of transcription.
- Hundreds of different transcription activators have been identified—each recognizes, and binds to, a specific UAS.
- Production of activators is triggered by:
  - environmental signals, e.g. a temperature rise that stimulates production of heat shock proteins
  - internal signals, such as hormones or growth factors; steroid hormones enter cells and bind to receptor proteins that then activate transcription through binding to a UAS—peptides cannot cross the cell membrane so trigger a signalling cascade.

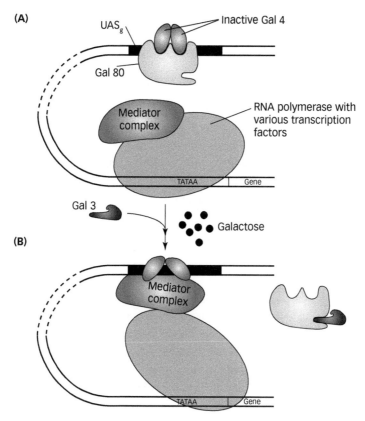

Figure 10.6 Activation of Gal 4 by galactose. (A) No galactose present: Gal 80 is bound to Gal 4; transcription repression of genes encoding galactose-metabolizing enzymes. (B) Galactose present: Gal 80 repression of Gal 4 removed; Gal 4 binds to, and activates, mediator complexes of genes to be transcribed.

- **Gal 4** is a much-studied transcription activator. It:
  - activates each gene of the enzymatic pathway that converts galactose to glucose-1-phosphate so that it can be used as a substrate for glycolysis
  - has become a model system for understanding how transcription activators work.
- Gal 4 activation occurs in the following way:
  - each of the enzyme-encoding *Gal* genes has a UAS sequence (UAS$_g$) to which Gal 4 binds
  - when galactose is absent, Gal 4 is bound to its UAS$_g$, but does not initiate transcription as inhibitory **Gal 80** protein is bound to Gal 4 (Figure 10.6)
  - when galactose is present another regulator protein, **Gal 3**, binds to, and stimulates, a conformational change in Gal 80. This binding frees Gal 80 from Gal 4, which now activates transcription of genes encoding galactose-metabolizing enzymes.

Control of gene expression in eukaryotes

- Gal 4, with its UAS, is often used experimentally to study gene expression in a range of eukaryotic species. Various cell lines are created containing the *Gal 4* gene and (UAS$_g$) next to the gene of interest.

(ii) Modifying chromatin structure

- An important part of regulating gene expression is the relaxation of the tight binding of DNA to nucleosomes and their compaction into 30-nm fibres, so that regulatory proteins and transcription factors can access DNA. Three processes are key to the remodelling of chromatin for transcription (Figure 10.7):
  - chemical modification of histone proteins
  - the repositioning of nucleosomes on DNA
  - levels of DNA methylation.

Histone modification

- The positively charged tail of histones is modified by the addition of phosphate, methyl, or acetyl groups.
- These modifications weaken the histone tails' interactions with DNA, enabling regulatory proteins and transcription factors to access DNA.
- Specific amino acids within the histone tails are preferentially modified. For example:
  - phosphorylation of serine residue 10 and acetylation of lysine residue 14 on histone H3 correlate with active transcription
  - methylation of lysine 4 in the tail of histone H3 is common at promoters of transcriptionally-active genes
  - acetylation of lysine 16 in the tail of histone H4 prevents formation of the 30-nm chromatin fibre; thus, chromatin is unpacked and available for transcription.
- For example, flowering in *Arabidopsis thaliana* is controlled by histone tail acetylation.
  - Flowering locus C (*FLC*) produces a repressor of flowering.
  - Flowering locus D (*FLD*) initiates floral development. It produces a deacetylase enzyme, which removes acetyl groups from histone tails of nucleosomes in the chromatin surrounding gene *FLC*. As a result, the chromatin condenses and *FLC* ceases transcribing its repressor of flowering genes.
- The pattern of histone modification (addition or removal of phosphate, acetyl, and methyl groups) that alters chromatin structure and regulates transcription is termed the **histone code**.

Nucleosome repositioning

- Chromatin remodelling involves repositioning nucleosomes so that regulatory proteins and transcription factors have access to relevant DNA sequences.
- At the core of chromatin remodelling complexes is an ATPase, which hydrolyses ATP to provide energy for nucleosome translocation.
- The best-studied example is the SW1/SNF complex found in yeast, humans, *Drosophila,* and other species.

DNA methylation
- Levels of DNA methylation determine the transcriptional activity of genes.
- Heavily methylated genes are not transcribed.
- Cytosine bases adjacent to guanine nucleotides are methylated: 5'-CpG-3'.
  ⊙ *See section 7.1 (p. 128) for details of DNA methylation.*
- Note that DNA methylation is distinct from methylation of histones.
- Methylated DNA attracts methyl-CpG-binding domain proteins (MBDs), which, in turn, bind histone deacetylases and other chromatin remodelling proteins that modify histones so that chromatin is compact and inaccessible for transcription.
- DNA regions with many CpG sequences are termed **CpG islands** and are common near transcription start sites.
- The predominantly non-methylated CpG islands in promoter regions of constitutively expressed housekeeping genes ensure their continuous transcription.
- In contrast, methylated CpG islands attract a complex of repressor proteins, which ensure transcription is inhibited.
- This link between DNA methylation and chromatin structure is key to determining which genes are expressed during development and differentiation.
- The human progressive neurodevelopmental disorder Rett syndrome is associated with disrupted DNA methylation patterns. Methyl-CpG-binding protein 2 (MeCP2), particularly prominent in brain neurones, is present at reduced levels in individuals with Rett syndrome.

Looking for extra marks?

The expression levels of thousands of genes can be monitored simultaneously by DNA microarray analysis. The **hybridization** of extracted cellular mRNA to tens of thousands of genomic DNA fragments, attached to a solid surface (e.g. a glass slide), is monitored. Microarrays are ideal tools to investigate differential gene expression during development, disease, pathogenic attack, or drug treatment.

 ### Check your understanding

10.2 Compare and contrast differences in gene organization in prokaryotes and eukaryotes, and how this affects mechanisms regulating initiation of transcription.

Post-transcriptional control

- The main post-transcriptional regulatory mechanism is the controlling of exon splicing from the pre-mRNA transcript.
- There are mechanisms that:

    (i) vary the pattern of introns removed—**alternative splicing**

    (ii) monitor splicing to ensure that all appropriate introns have been removed.

- There are numerous mechanisms that subsequently regulate cytoplasmic levels of mRNA and access of the translational machinery to mRNA. These are considered in the following 'translational control' section.

Alternative RNA splicing

- A single pre-mRNA is often spliced in alternative ways to produce different mRNAs (Table 10.3).
- Thus, different proteins (termed **isoforms**) can be made from a single gene.
- Isoforms tend to be tissue specific, e.g. the human thyroid hormone calcitonin and brain calcitonin gene-related peptide are products of the same gene. The pre-mRNA is processed differently in the thyroid gland and brain.
- Alternative splicing is ubiquitous in eukaryotes. A substantial proportion of higher eukaryotic genes produce multiple proteins in this way.
- Pre-mRNA from more than 90% of human genes is alternatively spliced. This helps to explain both the relatively low number of human protein-coding genes (the current estimate is ~21,000) and the great complexity of the human body.
- Cells also produce different isoforms at different stages of cellular differentiation. For example, the splicing pattern of vascular endothelial growth factor pre-mRNA varies at different developmental stages.
- When different splicing possibilities exist at several positions in the transcript, a single gene can produce dozens of different proteins.
- An extreme example is the **Dscam** (Down syndrome cell adhesion molecule) protein of *Drosophila melanogaster,* involved in migration and connection of neurons during embryogenesis.
  - There are 38,016 different versions of *Dscam*, depending on the splicing pattern of 4 of the 24 exons.
  - Exons 4, 6, 9, and 17 can be spliced in 12, 48, 33, and 2 different ways respectively.
  - This variable splicing seems pivotal for neuronal function: each fly neuron has a unique set of membrane Dscam proteins.
- Alternative splicing occurs in five main ways (Table 10.3); exon skipping is the most common mode in mammalian pre-mRNAs.
- The production of alternatively spliced mRNAs is regulated by a system of proteins that variously bind to sites on the pre-mRNA itself.
- **Splicing activators** promote the use of a particular splice site, while **splicing repressors** reduce a splice site's use.

Regulated nuclear transport

- 'Quality control' of mRNAs occurs in the nucleus.
- Incompletely processed mRNA never leaves the nucleus. Instead, it is degraded by RNA exonucleases located in **exosomes**—large multi-protein RNase-containing complexes found in both the nucleus and cytoplasm.

Type	Description	Example
Exon skipping	A particular exon may be retained or spliced	• Retaining or splicing exon 4 of the 6-exon *dsx* transcript is pivotal in sex determination in *Drosophila* • The transcription regulatory protein produced from mRNA containing exons 1, 2, 3, and 4 triggers female development • When the mRNA contains exons 1, 2, 3, 5, and 6, males develop
Mutually exclusive exons	One of two mutually exclusive exons is retained	• There are two mutually exclusive versions of exon 5 in the gene encoding human alpha tropomyosin • One version (SK) is used in skeletal muscle • The alternative exon 5 (NM) is used in non-muscle cells
Intron retention	A particular intron may be spliced or retained Intron	• Mouse transcription repressor *Tgif2* has two alternative forms depending upon splicing or retention of a 39 codon intron
Alternative 3′ splice sites	An alternative 3′ splice junction (acceptor site) is used, changing the 5′ boundary of the downstream exon	• *Caenorhabditis elegans* fibroblast growth factor receptor EGL-15 mediates different responses depending on the isoform produced by alternative 3′ splicing of exon 5
Alternative 5′ splice sites	An alternative 5′ splice junction (donor site) is used, changing the 3′ boundary of the upstream exon	• Exon 6 of the antibody heavy chain contains an internal splicer donor site • When the internal 5′ splice site is used the membrane-bound antibody results • 5′ splicing at the end of exon 6 produces the secreted antibody

Table 10.3 The five main patterns of alternative pre-mRNA splicing

• Transport of mRNAs (and other RNAs) through nuclear pores to the cytosol is directed by a family of **exportin** proteins.

Looking for extra marks?

Wilms tumour is a childhood cancer of the kidney. Patients often have a mutation in the *WT1* gene, which encodes a transcription factor regulating genitourinary development. One cause of Wilms tumour is a *WT1* splice mutation that disrupts alternative splicing at the exon 9 5′ splice donor site, leading to an imbalance of the transcription factor isoforms.

Check your understanding

10.3 How can multiple mRNAs be produced from one pre-mRNA?

Control of gene expression in eukaryotes

Translational control

- There is a variety of ways of regulating gene expression through controlling availability of the translational apparatus and its access to mRNA.
- These mechanisms, which control cellular levels of the encoded proteins, include regulating:
  - rates of RNA degradation
  - the binding of translational repressors to mRNA
  - subcelluar localization of mRNA.
- In addition, components of the translational apparatus may be inactivated temporarily.
- Sequences in the 3′ UTR of mRNA frequently mediate regulation of translation.

RNA interference

- Two types of small RNA molecules are important post-transcriptional/translational regulators of gene expression (Figure 10.7):
  - short interfering RNA (**siRNA**), which mediates cleavage and subsequent degradation of mRNA
  - microRNA (**miRNA**), which binds to mRNA and represses translation.
- Interaction between siRNA/miRNA and mRNA occurs in distinct regions of the cytoplasm termed **processing bodies** (**P bodies**).
- Both siRNA and miRNA are produced as longer double-stranded RNA molecules. This RNA migrates from the nucleus to cytoplasmic P bodies where:
  - the precusor dsRNA is cleaved by an endo-ribonuclease called **dicer** into 21–25 nucleotide segments
  - these short duplexes pair with proteins forming RNA-induced protein complexes (**RISC**)
  - one of the duplex RNA strands (the passenger strand) is degraded, while the other (the guide strand) becomes the functional siRNA or miRNA.
- siRNA and miRNA bind to complementary sequences within mRNA, generally within the 3′ UTR region:
  - siRNAs base-pair perfectly with their complementary sequences
  - it is rare for the match between miRNA and mRNA to be perfectly complementary.
- Regulation mechanisms by the two RNA types now differ:
  - the bound RISC/siRNA complex cleaves mRNA, which is then degraded
  - binding of miRNA to a target mRNA inhibits translation, which can then be activated at a future time.
- Translational silencing of oocyte mRNA by miRNA can last years!
- The discovery of RNA interference is a recent phenomenon: details are still being elucidated. Regulation of gene expression by RNA interference appears widespread in all eukaryotes, including humans.

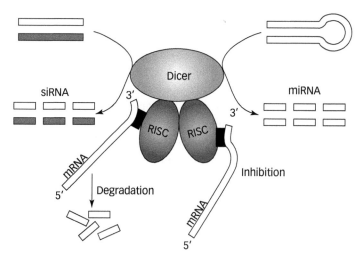

Figure 10.7 RNA interference.

Cellular localization of mRNA

- It is clearly advantageous to a cell for a particular mRNA to be located at the site where the protein product is required.
- Signals in the 3′ UTR of mRNA are often trigger localization mechanisms.
- Intermolecular base pairing within the 3′ UTR forms hairpin loops that are recognized by cellular proteins.
- For example, cytoskeleton **motor proteins** bind to 3′ UTR hairpin loops and use the energy from ATP hydrolysis to move the mRNA to specific locations where **anchor proteins** bind and retain the mRNAs.
- Other mRNAs diffuse randomly through the cytosol, and are trapped and retained by localized protein complexes. For example, actin-encoding mRNA is localized to the actin-rich outer cytosol regions, where actin is involved in cell surface movements.

Longevity of mRNA

- There is enormous variability in RNA longevity.
- Most mRNAs have a half-life of minutes to hours.
- Exceptions include mRNA in:
  - mammalian red blood cells, which digest their nuclei, but continue synthesizing haemoglobin and other proteins for several months, implying an mRNA lifespan of weeks
  - oocytes—mRNA is produced maternally, but not activated until fertilization.
- Lifespan information is encoded in the 3′ UTR of mRNA.
- AUUA in the 3′ UTR is the signal for ribonuclease degradation.
- Multiple copies of AUUA correlate with a short lifespan.

Control of gene expression in eukaryotes

- Some hormones protect against the action of ribonucleases.
- For example, the milk production hormone prolactin stabilizes casein mRNA and extends its lifespan; in lactating rat mammary glands, the lifespan of the mRNA is expanded from an average half-life of 1.1 h to 28.5 h.
- Rates of mRNA degradation are also controlled by sequences forming hairpin loops in the 3′ UTR of mRNA, e.g. degradation of transferrin receptor mRNA.
- Transferrin receptor imports iron into a cell:
  - when the cellular iron level is low, the enzyme aconitase is bound to the 3′ UTR hairpin loop of transferrin receptor mRNA, blocking its degradation
  - when the cellular iron level is high, aconitase binds iron, triggering its release from the 3′ hairpin loop; transferrin receptor mRNA is degraded.

Phosphorylation of translational initiation factors

- When a cell enters G_0, is infected by a virus, or lacks nutrition it reduces translation rates globally by phosphorylating eIF-2.
- eIF2/guanosine triphosphate (GTP) is associated with the methionine initiator tRNA.
- GTP is hydrolysed to guanosine diphosphate (GDP) as part of translation initiation.
- eIF-2b is required for dissociation of GDP from eIF-2.
- If eIF-2 is phosphorylated, eIF-2b binds irreversibly, blocking the recycling of GDP/GTP and so dramatically reducing translation.
- Phosphorylation of one of the eIF-4 subunits (eIF-4A) also inhibits translation in response to starvation.

  ➔ *See section 9.5 (p. 180) for details of translation initiation*

Post-translational control

- Post-translational use of a protein is regulated rigorously.
- Often the newly synthesized polypeptide is in an inactive form. Any of the following may then occur:
  - the cleavage of amino acid sequences from the nascent polypeptide
  - tertiary folding and internal bond formation
  - transport to specific subcellular locations
  - the addition of methyl, phosphates, and other chemical groups
  - association with other molecules, such as other proteins, lipids, sugars, and co-factors.
- For example, the pancreatic hormone insulin is synthesized in an inactive form, preproinsulin, which is processed to proinsulin and, finally, functional insulin (Figure 10.8):
  - the signal sequence is cleaved after the preproinsulin has been positioned by the ribosome in the endoplasmic reticulum

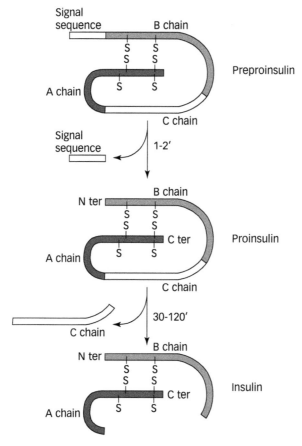

Figure 10.8 Post-translational processing of preproinsulin. N ter refers to the
N (or amino)-terminus; C ter refers to the C (or carboxyl)-terminus.

- ○ insulin chains A and B are then linked by internal disulphide bonds
- ○ an internal C peptide is cleaved once the insulin has been trafficked through
  the Golgi body into secretory granules; C peptide then becomes involved in
  microvascular blood flow.
- Enzymes frequently function in multi-step metabolic pathways, which are
  regulated by negative feedback inhibition.
- The pathway's end product binds to an allosteric site on the pathway's first
  enzyme, shutting down the pathway (Figure 10.9). This ensures metabolite supply
  meets demand.
- Post-translational modifications result in a **proteome** (all the proteins produced by
  a species) that is much larger than a species genome.
- For example, while it is currently estimated that the human genome contains
  ~21,000 genes, the proteome is estimated at over one million (Figure 10.10)!

Control of gene expression in eukaryotes

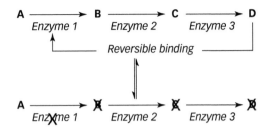

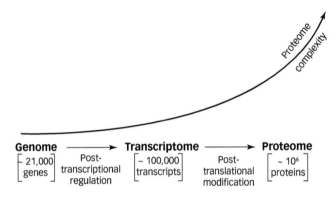

Figure 10.9 Allosteric regulation.

Figure 10.10 From human genome to proteome.

 Check your understanding

10.4 How do post-translation modifications to proteins regulate their activity?

10.5 What similarities and differences exist in prokaryotic and eukaryotic regulation of gene expression?

online resource centre You'll find guidance on answering the questions posed in this chapter—plus additional multiple-choice questions—in the Online Resource Centre accompanying this revision guide. Go to www. oxfordtextbooks.co.uk/orc/thrive/ or scan this image:

11 Gene Mutations

Gene mutations are heritable changes in the information encoded by a gene.

Key concepts

- Mutations create genetic variation. They are the only source of new variation within a genome.
- Mutations produce heritable changes in the genetic information.
- Generally, mutations have a detrimental effect.
- Mutation may be spontaneous or induced by mutagens in the environment.
- If mutations occur in somatic cells, they affect that individual only. If they occur in germ-line cells, they affect future generations.
- Mutations may be small scale, affecting a single gene, or on a larger scale affecting whole chromosomes.
- Cells have DNA repair mechanisms that can restore short sequences of damaged DNA to their original sequence.

11.1 TYPES OF GENE MUTATIONS

- Gene mutations alter the nucleotide sequence of a single gene.
- A gene mutation changes the DNA sequence of a single gene by:

Types of gene mutations

Mutation type	Original DNA sequence GGG ATC GAT GGA CAC	Codon reading frame consequence
Substitution	GGG ATT GAT GGA CAC	No change or single codon changed (encoded amino acid may/may not change; for details see Table 11.2)
Deletion	GGG AT*G ATG GAC AC	All codons (and amino acids) after the mutation changed
Insertion	GGG ATT TGA TGG ACA C	All codons (and amino acids) after the mutation changed

Table 11.1 Types of gene mutation
*Site of deletion.

- ○ substitution of a nucleotide
- ○ deletion or insertion of nucleotides.
- The change may or may not alter the amino acid sequence of an encoded protein (Table 11.1).

➡ *See section 9.1 (p. 172) for details of codons and the genetic code.*

Substitutions

- The most common type of gene mutation is the substitution of one nucleotide for another.
- Nucleotide substitutions are also referred to as base substitutions, as the change is often an *in situ* alteration to the nucleotide base.
- There are two categories of base substitution: **transitions** and **transversions** (Table 11.2).
- Although there are more possible transversions, transitions are more frequent because they are interchanges of bases of the same shape.
- The phenotypic consequences of a substitution on an encoded protein depend upon the site of mutation (Table 11.3).
- Substitution mutations produce the DNA sequence variants termed **single nucleotide polymorphisms** (SNPs).
- A substitution mutation can also negatively affect a protein if it occurs in a splice site sequence or a transcription regulatory sequence.

Transitions	Transversions
The substitution of a purine for a purine or of a pyrimidine for a pyrimidine	The substitution of a pyrimidine for a purine or of a purine for a pyrimidine
A → G C → T G → A T → C	A → C C → A A → T C → G G → C T → A G → T T → G

Table 11.2 Transition and transversion substitutions, and resulting possible base changes

Substitution type	Effect on encoded protein structure	Phenotypic consequence	Example
Silent	• Codon changed, but not amino acid • Explained by degeneracy of genetic code: multiple codons for some amino acids	• Generally none	• GTG → GTA • Valine encoded by both codons
Missense	• Codon change alters amino acid	• Minor/major (depends on specific change)	• GAG → GTG • Glutamine changed to valine • This change in β-globulin at amino acid 4 produces sickle cell anaemia
Nonsense	• Codon change produces a STOP codon • Truncated protein produced	• Major	• CGA → TGA • R1186X mutation of CFTR protein • Cystic fibrosis results

Table 11.3 The phenotypic consequences of a substitution mutation

Deletions and insertions

- The loss of one or more nucleotides from a gene is termed a **deletion**.
- **Insertion** refers to the addition of one or more nucleotides.
- Removing or inserting nucleotides into a gene generally changes the **reading frame** of all mRNA codons subsequent to the mutation (Table 11.1). Thus, deletions and insertion mutations are also termed **frameshift mutations**.
- Frameshift mutations generally have severe or lethal consequences.
  - For example, Duchenne muscular dystrophy is frequently caused by a frameshift mutation in the dystrophin gene.
- Insertions and deletions of multiples of three nucleotides do not alter the reading frame, but there can still be phenotypic consequences (Table 11.4).

Disease	Triplet repeat	Number of copies of triplet	
		Normal range	Disease range
Huntington disease	CAG	9–37	37–121
Spinal and bulbar muscular atrophy	CAG	11–33	40–62
Spinocerebellar ataxia	CAG	4–44	21–130
Fragile-X syndrome	CGG	6–54	50–1500
Myotonic dystrophy	CTG	5–37	44–1300
Friedrich ataxia	GAA	6–29	200–900

Table 11.4 Human diseases caused by insertions, i.e. by additional triplet repeats

Looking for extra marks?

Deletions and insertion mutations in the same gene at Xp21.2 cause the majority of cases of both Becker's and Duchenne muscular dystrophy (BMD and DMD). However, BMD is generally less severe, as the BMD-causing mutations tend to be in-frame, while the DMD causing mutations result in frameshifts.

Check your understanding

11.1 Why do mutations involving the insertion of a base usually have greater effects than those involving substitution of one base for another?

11.2 SPONTANEOUS MUTATIONS

- Mutations that occur as the result of natural processes in the cell are termed **spontaneous** mutations.
- They are generated by:
  - DNA replication errors
  - *in vivo* modification of base structure.

Replication errors

- DNA replication is extremely accurate.
- Uncorrected replication errors occur with a frequency of less than 1 in 10^9 nucleotides added.
- Replication errors result from:
  - mispairing of bases (A with C or G with T)
  - replication slippage.

Base mispairing
- Mispairing of bases is thought to occur because of base **tautomerization**, i.e. spontaneous changes in proton position in the bases (Figure 11.1):
  - the standard keto forms of G and T shift to rare enol configurations
  - the standard amino forms of A and C shift to rare imino configurations.
- A tautomeric shift produces a structural isomer with different bonding properties:
  - a modified purine pairs with the 'wrong' pyrimidine
  - a modified pyrimidine pairs with the 'wrong' purine.
- When a mispaired nucleotide is incorporated into a newly synthesized DNA strand, this is termed an **incorporated error** (T–G in Figure 11.1).
- During the subsequent round of DNA replication, an incorporated error becomes a permanent (transition) mutation if not corrected (C/G mutated to T/A; Figure 11.1).
- Most incorporation errors are detected and removed by the 'proofreading' 3′–5′ exonuclease activity of the replicating DNA polymerase.

  ➲ *See section 7.8 (p. 148) for details of DNA polymerase proofreading activity.*

Replication slippage
- Replication slippage is common in regions of repetitive sequence.

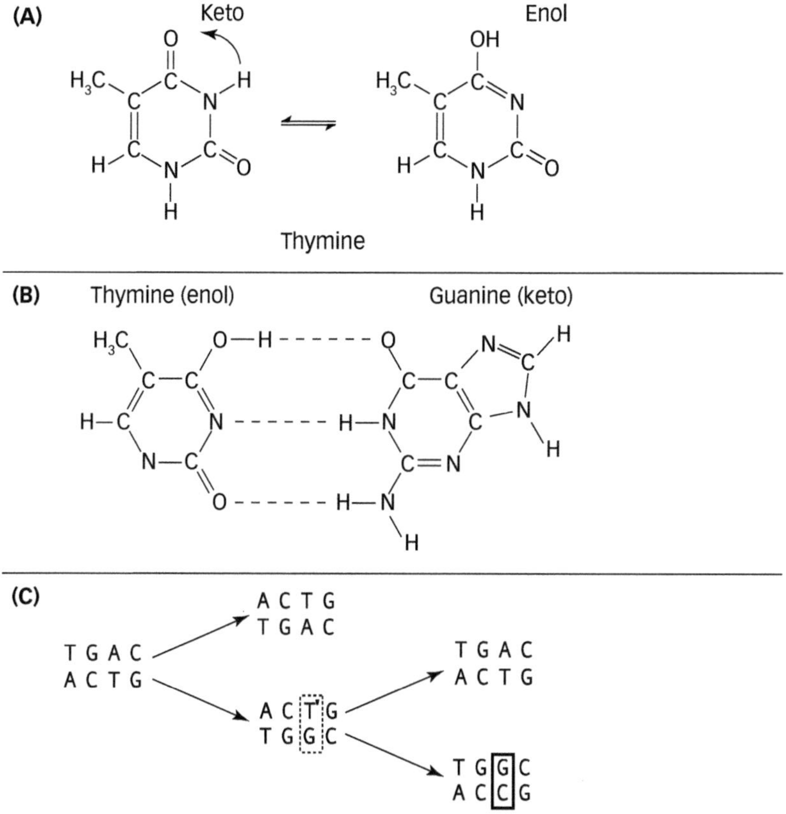

Figure 11.1 Base tautomerization leading to nucleotide substitution. (A) Arrow denotes a proton shift resulting in the rare enol isomer of thymine. (B) Enol isomer of thymine pairs with guanine during DNA replication (dotted lines represent hydrogen bonds). (C) Producing a transition mutation: a T base shifts to the enol form (T'), which mispairs with G, which then pairs with C in the subsequent round of replication.

- This refers to the 'slipping' of alignment of template and newly synthesized strands during DNA replication.
- As a consequence of slipping one of the DNA strands loops out. During the subsequent round of DNA replication these looped-out regions lead to insertions and deletions.
- Consider Table 11.5 carefully. If the looped out nucleotides were part of the newly synthesized strand an insertion mutation will occur; if they are part of the template strand this leads to a deletion mutation.
- Microsatellite polymorphism is mainly caused by replication slippage. If the mutation occurs in a coding region, it could produce abnormal proteins, leading to diseases. Huntington disease is a well-known human example resulting from triplet repeat expansion (Table 11.4).
  ➲ *See section 7.6 (p. 134) for details of microsatellites.*

Spontaneous mutations

An insertion mutation (CAT$_4$) to (CAT$_5$). Newly synthesized strand loops out. If loop not recognized and excised by DNA repair enzymes, it leads to an additional triplet.	A deletion mutation (CAT$_4$) to (CAT$_3$). Template strand loops out. If loop is recognized and excised by DNA repair enzymes, it leads to loss of a triplet.
** ATCGTAGTAGTA̲GTACCTG * TAGCATCATCATCATGGAC (loop GTA)	** ATCGTAGTAGTAGTACCTG * TAGCATCAT CATGGAC (loop CAT)

Table 11.5 Possible consequences of replication slippage involving one triplet (codon)
*Newly synthesized strand.
**Template strand.

Looking for extra marks?

DNA strand misalignment can also occur during homologous chromosome pairing at prophase I of meiosis. This causes unequal crossing over, which results in one DNA molecule with an insertion and the other with a deletion.

Base modifications

- DNA is damaged by oxidation, hydrolysis, or uncontrolled methylation of any of the four bases.
- The most frequent spontaneous damage is hydrolytic **depurination** and **deamination**.
- Depurination results in a loss of a purine base from a nucleotide. The covalent glycosidic bond joining a purine to the 1′ carbon atom of deoxyribose is broken, creating an **apurinic site**. **Apyrimidic sites** occur less frequently.
- Mutation results when DNA replication subsequently occurs. It is a two-stage process (Figure 11.2).
  - There is no base at the apurinic site for complementary base pairing. Thus, during the first replication any nucleotide (most frequently dATP) is incorporated opposite the apurinic site.
  - At the next round of replication the misincorporated nucleotide is used as a template. In Figure 11.2, thymine is incorporated opposite adenine. A substitution (transversion) mutation has occurred (G/C to T/A).
- Deamination (loss of an amine group) converts:
  - cytosine to uracil, which then pairs with adenine at the next replication (C/G → U/G → U/A → T/A)
  - adenine to hypoxanthine, which pairs with cytosine (A/T → Hyp/T → Hyp/C → G/C).
  - In both cases a substitution (transition) mutation has occurred.
  - Uracil in DNA is often recognized and removed (see section 11.3). Thus, a mutation is prevented.

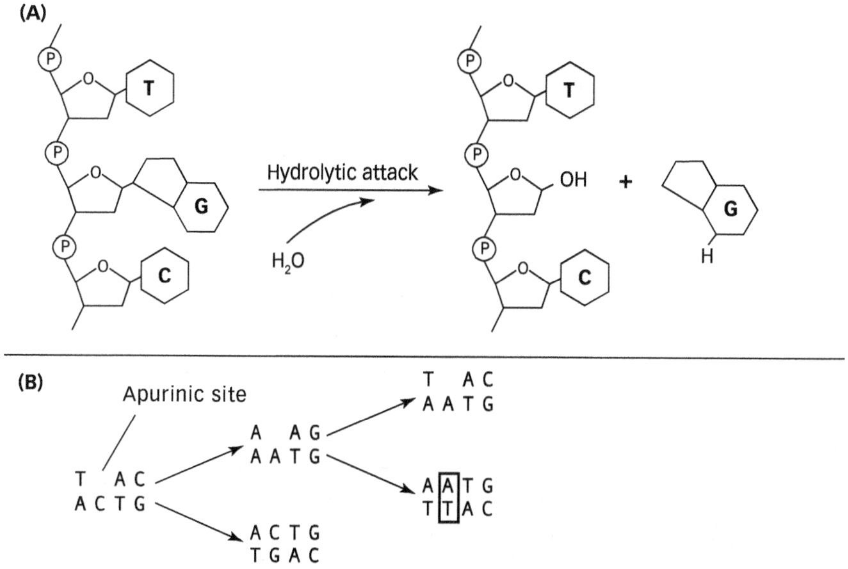

(A)

(B)

Figure 11.2 Transversion mutation resulting from depurination. (A) Hydrolytic attack removes a guanine. (B) Production of a transversion mutation.

- Many cytosine bases are methylated in DNA (section 7.1). Deamination of 5-methylcytosine forms thymine, a normal DNA base. Thus, 5mC/G → T/G → T/A transitions are frequent.
- Methylated CpG sites are termed **mutation hotspots**.

Looking for extra marks?

Hydrolytic depurination is one of the main forms of damage to ancient DNA, resulting in loss of base sequence information. Depurination can also be a problem for forensic samples.

Check your understanding

11.2 Draw a diagram to show the transition mutation A–T to G–C as the result of tautomeric shift in adenine.

11.3 INDUCED MUTATIONS

- Interaction of DNA with environmental agents, such as chemicals or radiation, produces **induced mutations**.
- Any outside agent that significantly increases the rate of mutation above the spontaneous rate is termed a **mutagen**.
- Transposable elements may also induce mutations during transposition.
  ⮕ *See section 7.7. (p. 135) for details of transposable elements.*

Induced mutations

Chemical induction

- Chemicals enter cells from the environment and cause mutations. Most of them are also carcinogens.
- Most chemical mutagens cause substitution mutations. They either:
  - chemically modify bases and change their pairing properties
  - are misincorporated during DNA replication and again alter pairing relationships.
- In both cases, when the next round of replication occurs a mutation is established.
- A few mutagens intercalate between base pairs causing insertions and deletions
- Table 11.6 summarizes the properties of the main classes of mutagens.

Class of mutagen	Example(s)	Mutagenic mechanism	Consequences
Base analogue chemical structure similar to A, T, C, or G	5- bromouracil	Analogue of T pairs with *both* A and G	GC → AT substitution (GC → GBU → ABU → AT)
	2-aminopurine	Analogue of A and G pairs with *both* T and C	CG → TA substitution (CG → CAP → TAP → TA)
Alkylating agent donates alkyl group (C_nH_{2n+1}) to base	Ethylmethylsulphate (EMS) CH_3-CH_2-O-SO_2-CH_3	Adds ethyl group to G forming ethylguanine which pairs with T	CG → TA substitution
Deaminating agents remove amine group from base	Nitrous acid HNO_2	Changes: C → U A → hypoxanthine G → xanthine	Mispairing of bases leading to substitution
Hydroxylamine (NH_2OH)		Hydroxylation: adds OH^- to C forming hydroxylamine cytosine, which pairs with A	CG → TA substitution
Free radicals	Superoxide radicals, H_2O_2	Modify base structures, e.g. oxidation of G → 8-oxy-7,8-dihydrodeoxyG	GC → AT substitution
Intercalating reagents	Acridine orange Proflavin Ethidium bromide	Positive charge attracts to DNA; wedge between adjacent bases; cause single nucleotide insertions and deletions	Frameshift mutations

Table 11.6 Chemical mutagens and their mode of action

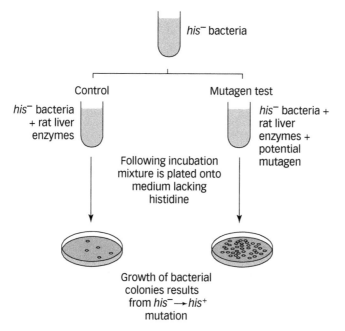

Figure 11.3 The Ames test. A solution of rat liver enzymes is incubated with the bacteria because liver enzymes sometimes convert a harmless compound into a mutagen.

- Various chemical mutagens are used in controlled genetic studies.
  - $\Rightarrow$ *See section 14.5 (p. 293) for the use of mutagens in genetic studies*
- The **Ames test** assesses the mutagenic potential of chemical compounds (Figure 11.3).
- It uses *his⁻* strains of the bacterium *Salmonella typhimurium* (i.e. a strain which cannot synthesize the amino acid histidine) and tests chemicals for their ability to induce a *his⁻* to *his⁺* mutation.
- To enhance the efficiency of the Ames test, the *his⁻* strains also have defects in:
  - the cell wall, so potential mutagens can easily enter the bacterial cells
  - **nucleotide excision repair** defects, thus preventing repair of DNA damage.
- As different mutagens act on DNA in different ways, the potential mutagen is incubated with several *his⁻* strains carrying different (substitution and frameshift) *his⁻* mutations.
- The Ames test is widely used to screen chemicals for their carcinogen potential, as most know carcinogens are also mutagens.

Looking for extra marks?

The common food additive sodium nitrate, also formed when meat is smoked, is converted to nitrous acid (NO_2) in the stomach. NO_2 deaminates C and A.

Radiation

- Both ionizing and non-ionizing radiation are mutagenic.
- High energy **ionizing radiation** (X-rays, and alpha, beta, and gamma rays) can damage DNA in two ways:
  - directly, by **double-strand breaks**, which lead to deletion of nucleotides and a variety of chromosomal rearrangements
  - indirectly, by knocking out electrons from atoms ('ionizing') to form free radicals, which modify bases; substitution mutations result.
- **Ultraviolet (UV) radiation** is less energetic, and therefore **non-ionizing**, but its wavelengths are preferentially absorbed by DNA bases.
- UV radiation induces **pyrimidine dimers**, i.e. bonds form between adjacent pyrimidine molecules on the *same* strand of DNA. **Thymine dimers** are most frequent, but cytosine dimers and cytosine–thymine dimers also form.
- Pyrimidine dimers:
  - disrupt hydrogen bonding between complementary bases
  - distort the DNA helix, which blocks DNA replication.
- If pyrimidine dimers are not repaired, cell division is inhibited and the cell dies.

 Check your understanding

11.3 What are the major causes and types of DNA damage?

11.4 DNA REPAIR SYSTEMS

- Mutation rates are low as cells possess mechanisms to:
  - prevent DNA damage
  - repair damaged DNA before it leads to a mutation.
- There are various cellular repair systems (Table 11.7), depending upon the type of damage:
  - DNA damaged in one strand is repaired according to sequence information in the undamaged complementary strand

Repair system	Type of damage repaired
Direct	Pyrimidine dimers, methylated bases
Base excision repair	Abnormal and modified bases
Nucleotide excision repair (NER)	Sequences with physical distortions, e.g. pyrimidine dimers
Mismatch repair (MMR)	Replication errors: mispaired bases and DNA loops
Homologous recombination (HR)	Double-strand breaks
Non-homologous end joining (NHEJ)	Double-strand breaks

Table 11.7 Summary of DNA repair mechanisms

- double-strand DNA damage uses sequence information from an undamaged sister chromatid or homologous chromosome.
- It is important to distinguish between DNA damage and DNA mutation.
  - 'DNA damage' refers to physical changes in the DNA, such as single- and double-strand breaks, and altered bases. These can be recognized by enzymes and correctly repaired if an undamaged complementary sequence is present.
  - A mutation is a permanent change in DNA sequence. Structurally, the DNA is normal. Thus, a mutation cannot be detected and repaired.
- Repair mechanisms are similar in prokaryotes and eukaryotes, and utilize a common four-step pathway:
  - detection—enzymatic recognition of DNA damage
  - excision—nucleases remove damaged nucleotide(s)
  - polymerization—DNA polymerase adds nucleotides to exposed 3'-OH group using an undamaged strand as the template
  - ligation—DNA ligase seals final nick in phosphodiester backbone.
- Failure to repair DNA leads to mutation, and is implicated in mechanisms of ageing and cancer development.

Preventing DNA damage

- Preventative mechanisms involve various cellular systems that recognize specific mutagens and detoxify them before they can damage DNA.
- For example, superoxide dismutase recognizes superoxide free radicals (O_2^-) formed from water by ionizing radiation and converts them to hydrogen peroxide, which is then detoxified by catalase:

$$2O_2^- + 2H^+ \quad \xrightarrow{\text{superoxide dismutase}} \quad H_2O_2 + O_2$$

$$2\,H_2O_2 \quad \xrightarrow{\text{catalase}} \quad 2H_2O + O_2$$

Direct reversal of DNA damage

- Cells can chemically reverse a limited amount of DNA damage.
- **Photoreactivation** occurs in bacteria and some eukaryotic cells (invertebrates and plants):
  - the enzyme **photolyase** directly repairs pyrimidine dimers
  - photolyase uses energy absorbed from the blue end of the visible spectrum (300–500 nm) to break the covalent cyclobutane bond that links pyrimidines in a dimer
  - normal cross-strand base pairing is restored (Figure 11.4).

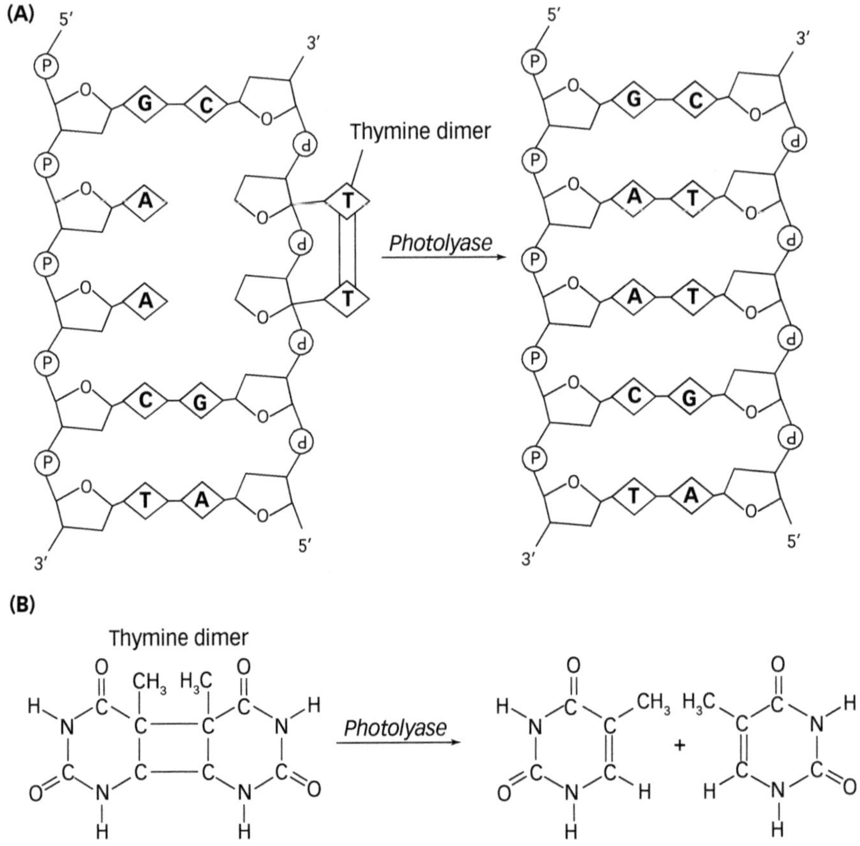

DNA repair systems

(A)

(B)

Figure 11.4 Direct repair of UV dimers. (A) *In situ* representation of thymine dimer repair. (B) Photolyase action.

- Methylation of guanine bases is directly reversed by methyl guanine methyl transferase (MGMT). MGMT becomes methylated so can only be used once.
- Methylation of cytosine and adenine can also be directly reversed in some species.
- Simple breaks in one strand are rapidly repaired by DNA ligase.

Base excision repair

- During **base excision repair** a small section of DNA that contains a modified (deaminated or depurinated) base is replaced (Figure 11.5).

Nucleotide excision repair

- Nucleotide excision repair (NER) recognizes and repairs many types of DNA damage.

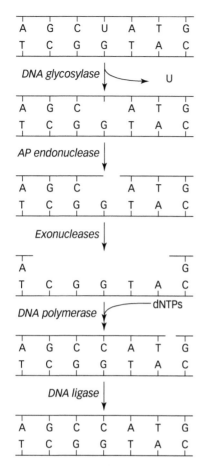

Details of Process

- A DNA glycosylase recognizes and removes the altered base.

- There are a set of different DNA glycosylases, each recognizing and removing a specific type of damaged base.

- For example uracil glycosylase recognizes and removes uracil produced by deamination of cytosine

- DNA glycosylases cut the glyscosyl linkage of the base to the 1' carbon of the deoxy-ribose

- Removal of the base produces an apurinic or an apyrimidinic (AP) site.

- An AP endonuclease cleaves the sugar and creates a nick in the sugar-phosphate backbone

- Exonucleases remove nucleotides near the nick, creating a gap

- The resulting gap is filled by a DNA polymerase, which adds new correct complementary nucleotides.

- The final nick in the sugar-phosphate backbone is sealed by DNA ligase; thus restoring the original sequence

Figure 11.5 Base excision repair.

- The trigger to initiate NER is damage that has physically distorted the DNA structure, e.g. pyrimidine dimers produced by UV light (Figure 11.6).
- In all cells a complex of enzymes scans DNA looking for distortions of its regular three-dimensional conformation.
- If a distortion is detected additional enzymes are recruited that separate the two nucleotide strands at the damaged region. Single-strand binding proteins stabilize the two single strands.
- DNA repair endonucleases cleave the phosphodiester backbone on both sides of the damage and break down the damaged DNA segment.
- The single-strand gap is filled by a DNA polymerase and DNA ligase seals the final nick in the sugar–phosphate backbone (Figure 11.6).
- There are two main classes of NER:
   ○ **global genome NER** (GG-NER), which repairs damage throughout the genome in both transcribed and non-transcribed DNA strands in active and inactive genes

DNA repair systems

- ○ **transcription-coupled NER** (TC-NER), which preferentially targets NER to the most important sequences, i.e. to repair damage in a DNA strand actively being transcribed.
- TC-NER is triggered when an RNA polymerase stalls at a DNA lesion.
- The blocked RNA polymerase is recognized by two proteins (CSA and CSB), which target the NER complex to the damaged site.
- Subsequent steps in TC-NER are the same as for GG-NER.
- Insights into the two NER mechanisms have come from the study of two human inherited diseases caused by defects in the DNA repair of UV-induced DNA lesions.
- **Xeroderma pigmentosum** (XP) and **Cockayne syndrome** (CS) are autosomal recessive skin disorders that result in enhanced sensitivity to sunlight. XP is associated with a predisposition to cancer, while neurological development is impaired in CS patients.
- The cells of people with XP are deficient in GG-NER owing to mutation in one of eight genes—XPA to G and XPV.
- Cockayne syndrome is associated with mutations in two genes (*CSA* and *CSB*), which promote TC-NER. The genes encode proteins that bind RNA polymerase and DNA respectively.

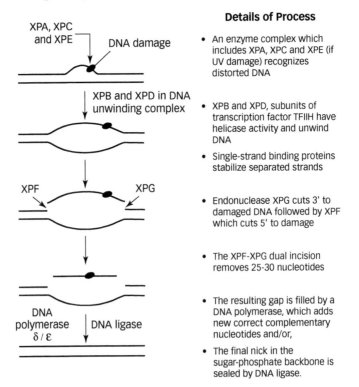

Details of Process

- An enzyme complex which includes XPA, XPC and XPE (if UV damage) recognizes distorted DNA

- XPB and XPD, subunits of transcription factor TFIIH have helicase activity and unwind DNA

- Single-strand binding proteins stabilize separated strands

- Endonuclease XPG cuts 3' to damaged DNA followed by XPF which cuts 5' to damage

- The XPF-XPG dual incision removes 25-30 nucleotides

- The resulting gap is filled by a DNA polymerase, which adds new correct complementary nucleotides and/or,

- The final nick in the sugar-phosphate backbone is sealed by DNA ligase.

Figure 11.6 Principle of nucleotide excision repair in eukaryotes, indicating roles of genes *XPA–G*.

- XP proteins A, C, D, and G have been shown to be homologues of *Escherichia coli* Uvr proteins **UvrA, UvrB, UvrC,** and **UvrD**, which are involved in NER.

Looking for extra marks?

The eighth XP gene, *XPV*, encodes DNA polymerase η. In addition to its role in nucleotide excision repair, DNA polymerase η can accurately replicate 8-oxoguanine produced by oxygen free radicals and also pyrimidine dimers.

Mismatch repair

- DNA replication is extremely accurate. Most incorrectly incorporated nucleotides are detected and excised by the proofreading capacity of DNA polymerase.
- This leaves ~1 in 10^7 misincorporated nucleotides which are corrected by **mismatch repair (MMR)**, also termed **post-replication repair**.
- Mismatch repair enzymes detect and correct misincorporated nucleotides and small unpaired DNA loops caused by strand slippage during DNA replication (see Table 11.5).
- After the replication error has been detected and the sugar–phosphate backbone nicked, exonucleases cut out a section of the newly synthesized strand.
- DNA polymerases fill the single-strand gap using the original strand as a template.
- Mismatch repair requires a way to distinguish between newly synthesized and original DNA strands to identify which nucleotide in a mismatched pair is the recently misincorporated one.
- In prokaryotes there is a **methylation signal**: the newly synthesized strand is unmethylated.
- Normally, adenine nucleotides in the prokaryotic sequence 5′GATC are methylated.
- There is a lag after replication before CH_3 groups are added to the new strand. Thus, the mismatch repair enzymes excise and replace nucleotides in the unmethylated strand.
- It is not known how the old and new strands are recognized in eukaryotes.
- One hypothesis is that the Okazaki fragments are a signal to the eukaryotic mismatch repair machinery of the nascent strand.
- In *E. coli* **mut proteins** (mut S, H, and L) detect mismatches and direct repair enzymes to the unmethylated strand (Figure 11.7).
- **Hypermutable** strains of *E. coli* result when mut *S*, *H*, or *L* are mutated.
- Homologues of *mut L* and *S* have been found in yeast and mammals.
- In humans mutation *mut L* and *mut S* homologues (termed *MLH1* and *MSH2* respectively) are associated with hereditary polyposis colorectal cancer.

DNA repair systems

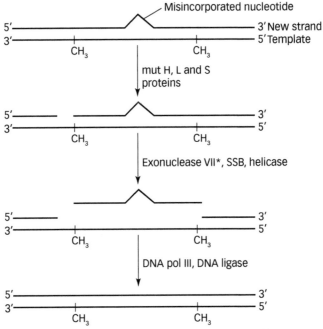

Figure 11.7 Mismatch repair in prokaryotes.

Homologous recombination

- Ionizing radiation, oxidative free radicals, and other agents cause double-strand breaks to DNA.
- Double-strand breaks are particularly detrimental to a cell as they:
  - stall DNA replication
  - lead to chromosomal rearrangements, such as deletions, inversions, duplications, and translocations.

  ➲ See section 12.3 (p. 236) for details of chromosomal rearrangements.
- Homologous recombination uses information from the DNA of the undamaged sister chromatid to restore the original DNA sequence at the site of the double-strand break.
- However, it can only occur in the S or G2 phases of the cell cycle, after DNA replication has occurred and while the two sister chromatids are held together at the centromere.
- Homologous recombination involves the following key stages (Figure 11.8):
  - (i) **Resection**—an exonuclease degrades the 5′ ends of the two broken strands, leaving a 3′-ended single strand that is used to prime DNA synthesis, which repairs the break

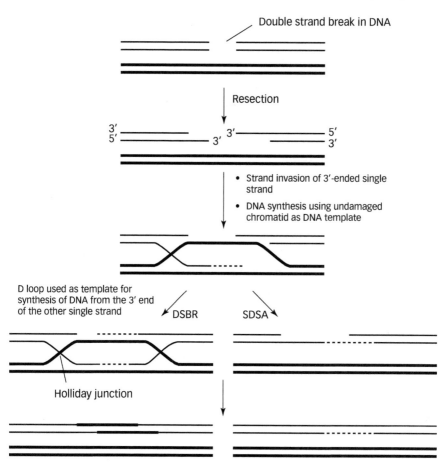

Figure 11.8 Principles of homologous recombination. Double-strand break repair (DSBR) and synthesis-dependent strand annealing (SDSA) are two ways of repairing a double-strand break if a sister chromatid or homologous chromosome is available.

(ii)   **Invasion**—the 3′-ended single strand invades the undamaged sister chromatid and anneals to its complementary strand of the sister chromatid

(iii) **Displacement**—the single-strand invasion displaces a strand of the unbroken DNA molecule; the displaced strand forms a D loop

(iv)   **Synthesis**—DNA synthesis occurs from both the 3′ end of the invaded single strand and from the 3′ end of the other broken strand, using the displaced D loop as a template.

- Following invasion of the single strand and formation of a D loop, **Holliday junctions** are established. These may be:
  ○ resolved immediately (synthesis-dependent strand annealing; Figure 11.8) and the second strand repaired using the re-annealed strand as template

DNA repair systems

Protein	Function
rec A (rad 51, eukaryotes)	• Critical recombination protein • Coats single-stranded DNA • Promotes single-strand invasion of a DNA helix and formation of a D loop
rec BCD	• Its helicase unwinds DNA • 5' to 3' exonuclease degrades on DNA strand • Remaining single strand invades DNA helix
Ruv AB	• A helicase that cataylses migration of D loop
Ruv C	• Endonuclease that cuts the two strands of DNA at a crossover (Holliday junction) • Resulting nicks joined by DNA ligase

Table 11.8 Key bacterial recombination proteins

- o where the D loop is used a template for repair synthesis of the second strand (double strand break repair; Figure 11.8).
- During crossing over at prophase I of meiosis homologous recombination occurs between non-sister chromatids of a pair of homologous chromosomes via the same mechanism as shown in Figure 11.8.
- Cross over points (Holliday junctions) are generated where a 3'-ended single strand invades and displaces the D loop. If endonucleases nick the DNA at the Holliday junctions, crossovers result.
- More than twenty-five different enzymes are involved in homologous recombination (see Table 11.8 for functions of four key enzymes).
- Bacterial mutants in recombination proteins are very sensitive to ionizing radiation and other agents causing double-strand breaks.
- Homologous recombination is faulty in human individuals with mutations in *rad51*, *BRCA1* and *2* (breast cancer susceptibility genes), and in Bloom syndrome.

Non-homologous end joining

- Non-homologous end joining repairs double-strand breaks without using a DNA template.
- It is used by cells in the G_1 phase of the cell cycle when a sister chromatid is unavailable for repair through homologous recombination.
- The broken DNA ends are simply brought together and rejoined by DNA ligase.
- Generally, there will be loss of nucleotides at the site of the double-strand break, so a deletion mutation occurs.
- As most of a higher eukaryotic genome is non-coding, rejoining broken chromosomes in this way is not detrimental to gene function.
- Instead, non-homologous end joining can be viewed as a quick and important repair mechanism that prevents further damage to DNA, e.g. by endonucleases degrading DNA from the cut ends.

SOS response

- The **SOS response** is a genomic global response triggered by major DNA damage.
- It is regulated in prokaryotes by two key proteins: Rec A and LexA.
- Repair genes are normally repressed by LexA.
- RecA-bound single-stranded DNA (resulting from DNA damage) activates RecA protease activity, which cleaves LexA.
- Loss of LexA induces transcription of SOS repair genes; for example **translesion DNA polymerase**, which is able to replicate past bulky DNA damage, such as a pyrimidine dimer.
- Any SOS repair is likely to be error prone, but can be corrected later.
- In eukaryotes, cell cycle checkpoints occur at the G_1/S and G_2/M boundaries (section 1.4). Kinases ATM and ATR respond to DNA damage and induce a signal transduction cascade that leads to cell cycle arrest.
- Depending upon the nature of the DNA damage and proteins activated, DNA repair or apoptosis results.

Check your understanding

11.4 Compare the two mechanisms available to cells to repair thymine dimers.

11.5 In what ways is mismatch repair more similar to nucleotide excision repair than base excision repair?

11.6 How do repair of double-strand breaks and single-strand lesions differ?

online resource centre You'll find guidance on answering the questions posed in this chapter—plus additional multiple-choice questions—in the Online Resource Centre accompanying this revision guide. Go to **www. oxfordtextbooks.co.uk/orc/thrive/** or scan this image:

12 Chromosome Mutations

Changes occur in the number and structure of chromosomes.

Key concepts

- One or a few chromosomes may be lost or gained.
- The addition of whole sets of chromosomes produces polyploid cells.
- Segments of individual chromosomes can be deleted, duplicated, become incorporated in other chromosomes, or inverted.
- Chromosome mutations often arise through errors during meiosis.
- In turn, chromosomal mutations frequently disrupt the process of meiosis, resulting in unbalanced gametes.
- Changes in chromosomal structure and number often occur in tumour cells.
- Chromosome mutations play important roles in evolution.

12.1 ANEUPLOIDY

- **Aneuploidy** refers to the loss or gain of one or a few individual chromosomes.
- Common aneuploid conditions are **monosomy** and **trisomy**. **Nullisomy** and tetrasomy also occur (Table 12.1).

Aneuploidy	Chromosomal explanation	Examples
Monosomy	2n–1 Loss of a single chromosome	• Human: Turner syndrome, loss of an X • *Drosophila*: monosomy chromosome IV
Trisomy	2n+1 Gain of a single chromosome	• Human: Down syndrome, extra chromosome 21 • Plants: trisomic series in Jimson weed, *Datura straminium*
Tetrasomy	2n+2 Gain of two homologous chromosomes (*not* any two extra)	• Human tetrasomy 18p (four copies of 18p; an extra chromosome composed of two copies of 18p)
Nullisomy	2n–2 Loss of both copies of a homologous pair of chromosomes	• Only occurs in allopolyploids (section 12.2), e.g. *Triticum aestivum* cv. Chinese Spring

Table 12.1 Types of aneuploidy

- Generally, aneuploidy has a major effect on an individual's phenotype (Table 12.2).
- Many aneuploids are lethal. For example, the only viable human monosomy is Turner syndrome (45,X). No human fetus survives if monosomic for an autosome.
- The detrimental consequences of aneuploidy result from abnormal gene dosage.
  - there is reduced gene product from a single monosomic chromosome
  - there are elevated levels of gene products in trisomic cells.
- Additional problems result in monosomic cells from the detrimental expression of recessive alleles from the single copy of a chromosome.
- Normally, in heterozygous diploid cells, deleterious recessive alleles are compensated for by the presence of a functional dominant allele.
- Aneuploidy results from errors during meiosis or mitosis.
- Most frequently, aneuploidy is the consequence of **non-disjunction** during meiosis. Meiotic non-disjunction may involve the failure of:
  - a pair of homologous chromosomes to separate at anaphase I
  - sister chromatids to separate at anaphase II.

Syndrome	Chromosomes	Frequency	Symptoms
Turner	X	1 in 2500	Mild; individual often has short stature and is infertile
Triple X	XXX	1 in 1000	Tall, learning problems, language defects, fertility unaffected
Klinefelter	XXY	1 in 600	Hypogonadism, infertility, weak secondary sexual development, learning problems
XYY	XYY	1 in 100	Tall, severe acne, some learning and behavioural problems
Patau	Trisomy 13	1 in 15,000	Multiple nervous and muscular complications; kidney and heart defects
Edward	Trisomy 18	1 in 8000	Kidney, heart and intestinal problems; microcephaly and distinctive facial characteristics
Down*	Trisomy 21	1 in 800	Learning problems, stunted growth, distinctive facial characteristics, congenital heart disease common

Table 12.2 Human aneuploids (other aneuploid fetuses generally fail to survive to term)
*~5% of Down syndrome results from Robertsonian translocation of chromosome 21 (see section 12.2).

Aneuploidy

- Non-disjunction produces gametes which either lack a chromosome or have an extra chromosome.
- Depending upon which gamete combines with normal haploid gametes, the resulting zygote is monosomic or trisomic. (Figure 12.1 shows formation of a trisomic zygote.)
- Non-disjunction occurs during a mitotic division if sister chromatids fail to separate (Figure 12.2).
- Somatic clones of monosomic and/or trisomic cells are produced.
- Mitotic non-disjunction early in embryonic development produces a **mosaic** individual who has patches of cells with different chromosome numbers.
- For example, about 50% of individuals diagnosed with Turner syndrome are mosaic, i.e. they have a mixture of 45,X and 46,XX cells.
- An aneuploid cell may also be produced during mitosis or meiosis if a centromere is deleted from a chromosome. Without a centromere there is no attachment

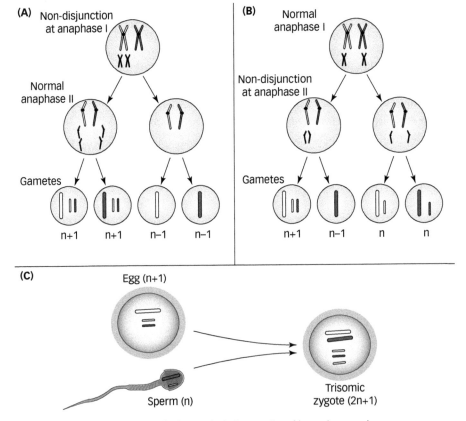

Figure 12.1 Non-disjunction during meiosis (two pairs of homologous chromosomes are shown). (A) Non-disjunction at anaphase I. (B) Non-disjunction at anaphase II. (C) Production of a trisomic zygote.

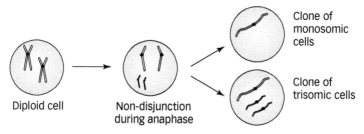

Figure 12.2 Non-disjunction during mitosis (one pair of homologous chromosomes shown).

point for spindle fibres and so ordered segregation of chromatids, or homologous chromosomes, to new nuclei is disrupted.

- Clones of aneuploid cells are associated with various cancers, e.g. trisomy 12 in chronic lymphocytic leukaemia and trisomy 8 in acute myeloid leukaemia.

Uniparental disomy

- **Uniparental disomy** (UPD) results when both copies of a pair of homologous chromosomes come from the same parent.
- Aneuploidy, followed by **trisomic rescue**, leads to uniparental disomy.
- Trisomic rescue occurs when a trisomic embryo loses one of its triplicate chromosomes early in embryonic development, thus restoring the normal diploid chromosome number.
- Uniparental disomy describes the situation when, by chance, the two remaining chromosomes following trisomic rescue are from the same parent (Figure 12.3).
- Uniparental disomy can disrupt gene expression. For example:
  - unexpected expression of recessive alleles when *only* one parent is a carrier (Figure 12.4)

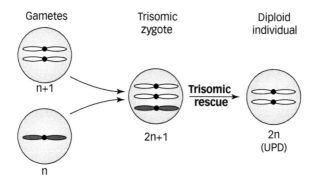

Figure 12.3 Uniparental disomy (UPD): trisomic rescue produces an individual with a pair of chromosomes from the same parent.

Polyploidy

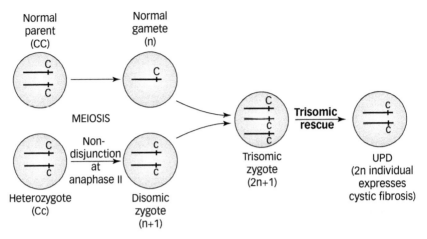

Figure 12.4 Uniparental disomy explains unexpected inheritance of cystic fibrosis: parents CC and Cc.

○ expression of imprinted genes, e.g. if an individual has two imprinted (methylated) maternal q11.2q13.1 regions of chromosome 15, then he/she expresses Prader–Willi syndrome.

➔ *See section 3.7 (p. 56) for details of imprinted genes.*

Looking for extra marks?

There are two types of UPD:
- **isodisomy**, i.e. when UPD is the result of non-disjunction during anaphase II. The two chromosomes in the resulting disomic gamete come from one homologue (as shown in Figure 12.4)
- **heterodisomy**, i.e. when non-disjunction occurs during anaphase I and the disomic gamete contains a chromatid from each homologue. In Figure 12.4, heterodisomy would have resulted in a heterozygous individual (Cc).

➔ Check your understanding

12.1 How are aneuploid individuals produced?

12.2 A karyotype was prepared for a newborn boy. He was found to be mosaic trisomy 8. How was this produced?

12.3 Females with five X chromosomes (49,XXXXX) have been reported. How could this arise?

12.2 POLYPLOIDY

- The presence of additional whole sets of chromosomes is called **polyploidy**.
- Generally, all cells of an individual are polyploid.

- If only certain tissues within an organism have cells with extra sets of chromosomes this is termed **endoploidy**.
- Polyploidy is common in plants. About half of all flowering plant species are polyploids.
- Polyploidy is a major mechanism in plant evolution. Polyploid plants often have a selective advantage as they:
  - are larger than diploid relatives (larger leaves, flowers, seeds)
  - grow more vigorously.
- The increased size of polyploids is explained by larger cells resulting from the increased nuclear volume needed to accommodate extra chromosomes.
- Because of the greater size and productivity of polyploid individuals, many commercially grown plants are deliberately-induced polyploids (Table 12.3).
- Polyploidy is much less common in animals, but found in invertebrates, fishes, salamanders, frogs, and lizards.
- One polyploid mammal is known: the red Visacha (4n) rat in Argentina.
- Polyploidy often produces fertility problems because extra chromosome sets:
  - disrupt sex-determining mechanisms so individuals are sterile
  - cause problems in the pairing of homologous chromosomes during prophase I of meiosis and subsequent segregation during anaphase I; thus, viable gametes are not produced.
- Plants can often overcome fertility problems by reproducing asexually, which explains the greater incidence of polyploidy in plants compared with animal species. Fewer means of asexual reproduction exist in animals.
- There are two types of polyploidy depending on the origins of the extra sets of chromosomes:
  - **autopolyploidy**—extra sets of chromosomes are derived from the same species, e.g. the autotetraploid AAAA, where A represents the haploid chromosome set
  - **allopolyploidy**—chromosomal sets come from different species, e.g. the allopolyploid AABB where A and B represent chromosome sets from two different species.

Ploidy	Examples
Triploid, 3n	Bananas, seedless watermelons, tardigrads (animals)
Tetraploid, 4n	Peanuts, cotton, Salmonidae (fish), red Visacha rat
Pentaploid, 5n	Kenai birch (tree)
Hexaploid, 6n	Wheat, kiwifruit
Octaploid, 8n	Strawberries, *Acipenser* (genus of sturgeon fish)
Dodecaploid, 12n	Some blackberry varieties, *Xenopus ruwenzoriensis* (amphibian)

Table 12.3 Examples of polyploids

Polyploidy

Autopolyploidy

- Autopolyploids can arise in three ways.
  - (i) **Meiotic failure.** If all homologous pairs of chromosomes fail to segregate during anaphase 1 *or* all pairs of sister chromatids fail to segregate during anaphase II of meiosis the result will be a diploid gamete. If this gamete fuses with a normal haploid gamete a triploid zygote, and so triploid individual, is formed.
  - (ii) **Mitotic failure.** Non-disjunction of all chromosomes during a mitosis early in embryonic development doubles the chromosome number and so produces an autotetraploid (4n).
  - (iii) **An unusual fertilization event.** Two sperm, or two pollen nuclei, may fertilize a single ovum (**dispermy**). This produces a triploid individual. One to two per cent of all human fetuses are triploid, most caused by dispermy. Triploid fetuses abort spontaneously.
- Autopolyploids rarely produce viable offspring.
- This infertility results from problems in the pairing of homologous chromosomes and their segregation during meiosis (Figure 12.5).
- Consider an autotriploid—no matter how the three homologues of one chromosome align during prophase I (all three, two, or none pairing) they will segregate randomly at anaphase I producing **unbalanced gametes**, i.e. gametes with variable number of chromosomes (some chromosomes may be missing and others present in multiple copies).
- The sterility of autoploids is exploited commercially. For example, wild diploid bananas (2n=22) contain hard, inedible seeds. Commercially grown bananas are autotriploids (3n=33). They are sterile and so fail to produce seeds.

Allopolyploidy

- An allopolyploid is produced in the following way:
  - hybridization occurs between two closely related species
  - the resulting hybrid is diploid and sterile—viable gametes cannot be produced as the maternal and paternal chromosome sets are from different species and so cannot pair during prophase I of meiosis
  - the hybrid is propagated asexually
  - eventually, the hybrid undergoes chromosome duplication; typically, mitotic non-disjunction occurs, e.g. in embryonic floral tissue producing fertile allopolyploid flowers
  - normal meiosis is now possible as chromosome doubling results in two separate homologous sets of chromosomes—the homologous chromosomes of each set pair and segregate properly, producing balanced gametes (Figure 12.6).
- Many of the world's commercially important crops are allopolyploids.

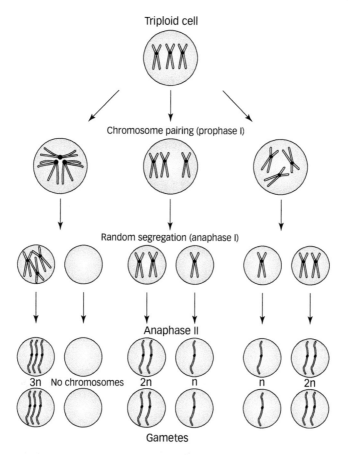

Figure 12.5 Meiosis in an autotriploid. The pairing and segregation of one homologous set of chromosomes is considered. (A) All three homologues pair; (B) two homologues pair; (C) no pairing occurs.

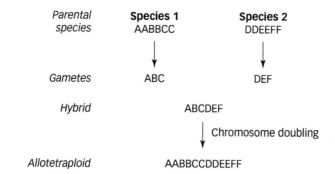

Figure 12.6 The production of an allotetraploid (4n=12). Species 1 (2n=6) with chromosome pairs AABBCC hybridizes with species 2 (2n=6) with chromosome pairs DDEEFF.

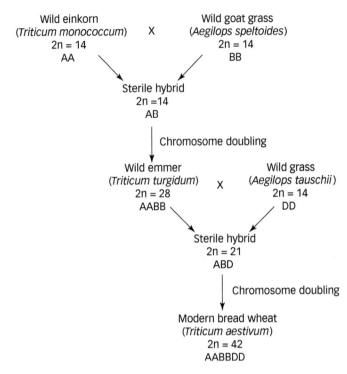

Figure 12.7 Recent evolutionary history of modern bread wheat (*Triticum aestivum*).

- Modern cultivated wheat is an allohexaploid (Figure 12.7). Two natural hybridization and chromosome doublings are believed to have occurred.
- Cultivated cotton is an artificially-induced allotetraploid between an Old World and a wild American species. Colchicine was used to induce chromosome doubling in the hybrid.

Check your understanding

12.4 Distinguish between aneuploidy, diploidy, haploidy, and polyploidy.
12.5 How does the occurrence of non-disjunction during meiosis produce individuals made up of cells with different numbers of chromosomes?

12.3 CHANGES IN CHROMOSOME STRUCTURE

- Chromosomal segments may be:
  - deleted
  - duplicated

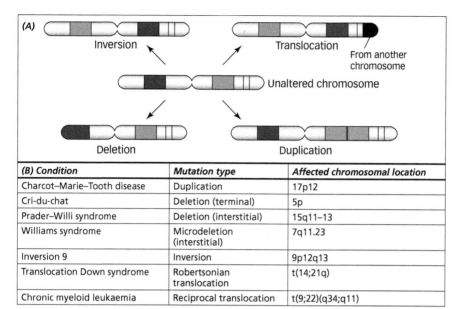

(B) Condition	Mutation type	Affected chromosomal location
Charcot–Marie–Tooth disease	Duplication	17p12
Cri-du-chat	Deletion (terminal)	5p
Prader–Willi syndrome	Deletion (interstitial)	15q11–13
Williams syndrome	Microdeletion (interstitial)	7q11.23
Inversion 9	Inversion	9p12q13
Translocation Down syndrome	Robertsonian translocation	t(14;21q)
Chronic myeloid leukaemia	Reciprocal translocation	t(9;22)(q34;q11)

Table 12.4 (A) Four main types of chromosomal rearrangements. (B) Examples of human chromosomal rearrangements

Terminal/interstitial refers to chromosomal location of deletion, at the end or within a chromosomal arm.

- ○ inverted
- ○ translocated.
- Alterations of a chromosome's structure result from breaks along its axis. These breaks can:
  - ○ occur spontaneously, e.g. chromosomal breakage is intrinsic to the process of crossing over during prophase I of meiosis
  - ○ be induced by chemicals or radiation.
  - ➔ *See section 11.2 (p. 212) for details of induced mutations.*
- A broken piece of chromosome readily joins with another broken piece. If joining does not re-establish the original relationship a chromosome mutation results (Table 12.4).
- There may be an immediate or delayed phenotypic consequence.
  - ○ Immediate effects result if, for example, a crucial gene has been deleted or the chromosomal breakage accompanying a translocation occurred within a vital gene, abolishing its function.
  - ○ Delayed effects result when there is no immediate disruption of gene function, but the rearrangement causes problems during the next meiosis, resulting in unbalanced gametes.
- Many chromosomal rearrangements cause problems for homologous pairing of chromosomes during prophase I (discussed in following sections).

Chromosomal duplications and deletions

- The most common cause for gain or loss of chromosomal material is **unequal crossing over** during prophase I of meiosis.
- During crossing over between non-sister chromatids, the two DNA molecules may misalign. This results in one DNA molecule with an insertion and the other with a deletion.
- Unequal crossing over occurs readily when homologous sequences are tandemly repeated (Figure 12.8).
- Chromosomal duplications and deletions generate similar problems:
  - gene dosage imbalances
  - unbalanced gametes.

Gene dosage imbalances

- Many cellular processes require a carefully regulated interaction of many proteins; thus, having the correct relative amounts of these different proteins is critical. This is particularly important in eukaryotic developmental processes.
- If the amount of one protein increases or decreases relative to that of others, problems can result, e.g. in multimer assemblage.
- A duplicated gene produces a cell with three functional genes and so 50% more gene product than the normal two copies.
- Duplicated *Bar* genes in *Drosophila melanogaster* result in small, bar-shaped eyes, which cause impaired vision (Table 12.5).
- **Haploinsufficiency** results from a deleted gene, i.e. the remaining copy of a gene does not provide sufficient gene product for normal function.
- Genetically, **pseudodominance** is often the consequence of the deletion of a chromosomal region.
- Pseudodominance refers to the expression of a normally recessive allele due to deletion of the section of chromosome with the dominant allele in an individual who was originally heterozygous (**Aa**).

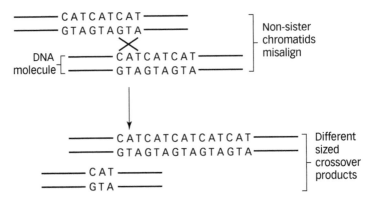

Figure 12.8 Unequal crossing over between two chromatids.

Genotype	Number of eye facets	Eye phenotype	X chromosome region 16A (showing bar genes)
Wild type female +/+	780		
Wild type male +	740		
Heterozygous female (B/+)	350		
Homozygous female (B/B)	70		
Mutant male (B)	90		

Table 12.5 X-linked gene duplication produces Bar phenotype in *Drosophila melanogaster*

- The genotype of an individual becomes **a** instead of **Aa**.
- Expression of the recessive allele often has detrimental effects.

Unbalanced gametes

- The pairing of homologous chromosomes during prophase I of meiosis is disrupted because of additional or missing chromosomal regions in one of the homologues.
- To compensate, the additional chromosomal material loops out during the pairing of homologues (Figure 12.9).
- Chromosomal loops are vulnerable to degradative enzyme attack, resulting in unbalanced gametes, i.e. gametes lacking chromosomal segments.

Looking for extra marks?

Gene duplication is believed to play a positive role in evolution.

- A duplicated gene is free from selective pressure, i.e. it can mutate without deleterious consequence to the organism.
- Eventually, it could code for a new function.
- Evidence for a role of gene duplication in evolution comes from pairs of genes which share a substantial part of their sequence and have similar function, e.g. those encoding protein digestive enzymes trypsin and chymotrysin, or the alpha and beta genes of haemoglobin.

Changes in chromosome structure

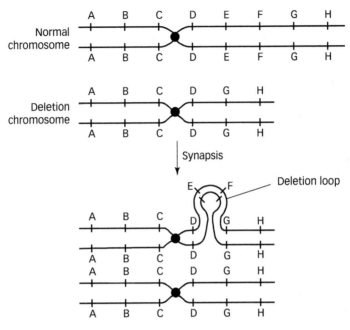

Figure 12.9 Meiosis in a deletion heterozygote. At synapsis during prophase I the unpaired chromosome segment loops (letters represent genes). Note that a similar structure would form during synapsis between a normal chromosome and one with a duplicated segment.

Chromosomal inversions

- Inversions arise when two breaks occur in a chromosome and the intervening segment is turned 180 degrees.
- There are two types of inversions (Table 12.6).
- **Inversion heterozygotes** (i.e. individuals with an inversion in one of a pair of homologous chromosomes) are common in some taxa, e.g. many flowering plant and insect species.
- Approximately 2% of humans carry a symptomless pericentric inversion of chromosome 9.

Paracentric inversion	*Pericentric inversion*
Paracentric inversions do not include the centromere. Both breaks occurred in one arm of the chromosome.	**Pericentric inversions** include the centromere. A breakpoint occurred on each chromosome arm.
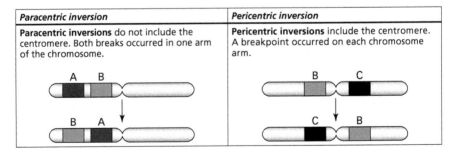	

Table 12.6 The two types of chromosomal inversions

- Lowered fertility is found in inversion heterozygotes because of meiotic problems.
  - During meiosis, in order for the relevant two homologous chromosomes to pair during prophase I, they form an **inversion loop** (Figure 12.10).
  - If crossing over occurs in an inversion loop, unbalanced gametes are produced because abnormal chromatids are generated. Some resulting chromatids have two copies of some genes, while others lack genes.
  - Furthermore, crossover in a paracentric inversion generates chromatids with two centromeres, termed **dicentric** chromatids, and others lacking a centromere—**acentric** chromatids.
  - If any of these gametes carrying these abnormal chromosomes are used in fertilization, the resulting zygotes are invariably unviable.
- Carefully study Figure 12.10 to work out the consequences of a crossover in a paracentric inversion loop.

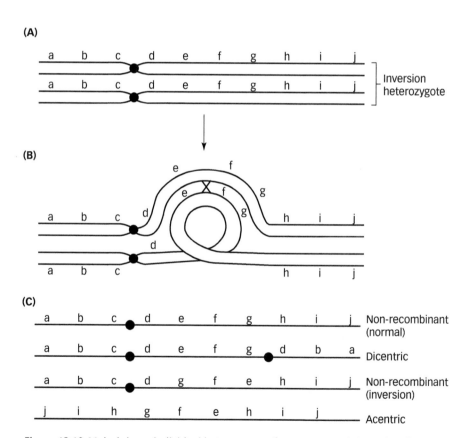

Figure 12.10 Meiosis in an individual heterozygous for a paracentric inversion (letters represent genes). (A) Chromosomes of the inversion heterozygote; (B) inversion loop that forms when homologous chromosomes pair during prophase I; (C) chromosomes that result from a crossover in the inversion loop.

Changes in chromosome structure

Chromosomal translocations

- There are two types of translocation: **reciprocal** and **Robertsonian.**
- A **reciprocal translocation** is more common. It involves a two-way exchange of material between non-homologous chromosomes.
- There is no net gain or loss of chromosomal material: two chromosomes have been broken and rejoined in the wrong combination (Figure 12.11).
- A Robertsonian translocation is produced by the fusion of two **acrocentric** chromosomes.
  - ○ The centromeres of the two chromosomes fuse, forming one large chromosome with two long arms.
  - ○ The short arms are lost. This results in an overall reduction in chromosome number, e.g. in humans, 45 instead of 46 chromosomes.
- In humans, Robertsonian translocations involve two of chromosomes 13, 14, 15, 21, and 22.

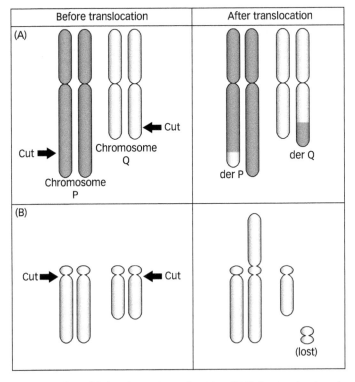

Figure 12.11 Formation of (A) reciprocal translocation (B) Robertsonian translocation. Note that 'der' refers to derivative chromosome, i.e. a chromosome produced as the result of a structural rearrangement.

- The short arms of human acrocentric chromosomes have tandem copies of ribosomal RNA genes. Losing two out of ten rRNA blocks does not cause genetic problems.
  ➔ *See section 1.2 (p. 3) for details of chromosome morphology.*
- Translocations occur during formation of the egg and sperm, or shortly after conception.
- Most reciprocal and Robertsonian translocations are **balanced**, with no severe phenotypic consequences for the carrier individual.
- There are, however, consequences for the next generation, as problems result during segregation of chromosomes during anaphase 1 of meiosis in an individual with a balanced translocation.
- Consider first a reciprocal translocation. There will be two pairs of homologous chromosomes variously sharing homology. Each chromosome has regions that are

(A)

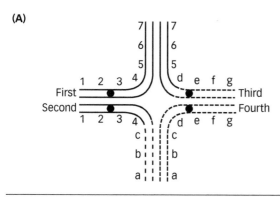

(B)

	Alternate segregation		Adjacent segregation 1		Adjacent segregation 2	
	First & fourth	Second & third	First & third	Second & fourth	First & second	Third & fourth
	1 a	1 g	1 g	1 a	1 1	g a
	2 b	2 f	2 f	2 b	2 2	f b
	3 c	3 e	3 e	3 c	3 3	c c
	4 d	4 d	4 d	4 d	4 4	d d
	5 e	c 5	5 5	c e	5 c	5 e
	6 f	b 6	6 6	b f	6 b	6 f
	7 g	a 7	7 7	a g	7 a	7 g

Normal   Reciprocal translocation (balanced)   All unbalanced gametes

Figure 12.12 Meiosis in an individual carrying a reciprocal translocation. (A) Tetravalent formed during prophase I. (B) Three segregation patterns and their products (continuous lines represent one homologous pair of chromosomes with genes 1–7; broken lines represent the other pair of chromosomes with genes a–g).

Changes in chromosome structure

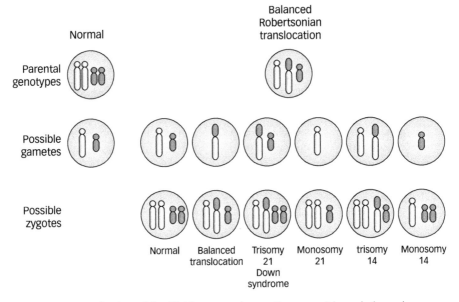

Figure 12.13 Production of familial Down syndrome. One parent has a balanced Robertsonian translocation involving chromosomes 14 and 21 (unshaded chromosome 14 and shaded chromosome 21).

homologous with two others. As a result, all four chromosomes associate during prophase 1 forming a **tetravalent** (Figure 12.12A).

- There is only one pattern of segregation—**alternate segregation**—at anaphase 1, which produces balanced viable gametes. The gametes produced by the other patterns of segregation are imbalanced: partially trisomic and monosomic for regions (study Figure 12.12B).

- There are similar pairing and segregation problems during meiosis I in an individual carrying a balanced Robertsonian translocation. In such cases the chromosomes pair in meiotic prophase to form a **trivalent**.

- Figure 12.13 considers a Robertsonian translocation between human chromosomes 14 and 21. One of the possible unbalanced gametes will contain, effectively, two copies of chromosome 21. Thus, an individual with this translocation is at risk of producing a baby with Down syndrome.

- About 5% of individuals with Down syndrome result from a parent carrying a balanced Robertsonian translocation.

Ring chromosomes and isochromosomes

- Ring chromosomes and isochromosomes may arise spontaneously or as the result of chromosomal damage by mutagens (Table 12.7).

- They are commonly found amid the complex chromosomal abnormalities formed in malignant tumours (section 12.4).

Description	Formation
• The p and q arms fuse into a **ring chromosome** • Normally, telomeres are lost • Rare, but ring chromosomes have been found for most human chromosomes • For example, ring chromosome 20 syndrome is associated with epilepsy and ring chromosome 14 with mental retardation	
• **Isochromosomes** are formed by loss of one arm and duplication of the remaining arm • There must be a centromere • Sometimes isochromosomes have two centromeres	

Table 12.7 Ring chromosomes and isochromosomes

Check your understanding

12.6 Who is likely to be more at risk of producing a child with Down syndrome, and why: a 23-year-old woman who carries a balanced 14/21 Robertsonian translocation or a 43-year-old who has no previous family history of Down syndrome?

12.4 CYTOGENETIC NOMENCLATURE

• The International System for Human Cytogenetic Nomenclature (ISCN) is a standard way of representing the different types of human chromosome abnormality (Tables 12.8 and 12.9).

• Chromosomal details of an individual are stated in the following order: number of chromosomes, sex, first chromosomal change, second chromosomal change, etc.

• Sex chromosome changes are listed first and then others in chromosomal numerical order.

• Details of structural changes are given in brackets, i.e. which chromosomes are affected and the location of the change within the chromosomes.

Cytogenetic nomenclature

Symbol	Description	Symbol	Description
–	Missing chromosome	del	Deletion
+	Additional chromosome	dup	Duplication
p	Short arm of a chromosome	inv	Inversion
q	Long arm of a chromosome	i	Isochromosome
ter	End part of chromosome arm	t	Translocation
()	Groupings for details of structural alterations	r	Ring chromosome
		der	Derivative chromosome

Table 12.8 Commonly used International System for Human Cytogenetic Nomenclature abbreviations

ISCN	Description
46,XX	Normal female
69,XXY	triploid male
47,XY,+18	Male with trisomy 18, Edward syndrome
47,XY,+8/46,XY	Male who is a mosaic of trisomy 8, Warkany syndrome
45,XX,der(13;14)(q10;q10)	Female with a Robertsonian translocation between chromosome 13 and 14; q10 indicates a centromere breakpoint
46,XY,del(7)(q11.23q21.2)	Male with a deletion on chromosome 7 between bands q11.23 and 21.2 on the long arm
46,XX,inv2(p21q13)	Female with a pericentric inversion (as the inversion extends from p21 to q13, it includes the centromere)

Table 12.9 Examples of International System for Human Cytogenetic Nomenclature for different karyotypes

- For example, 46,XY,der(4)t(4;8)(p16;q22) represents a male with a derivative chromosome 4 which resulted from a translocation between the short arm of chromosome 4 at band 16 and the long arm of chromosome 8 at band 22 (p16 and q22 state the chromosomal breakpoints where the reciprocal translocation occurred).
- The ISCN band numbering system refers to Giesma-stained bands in a normal karyotype (see section 1.2).
- Band numbering of a chromosome arm begins at its centromere and increases towards the telomere (Figure 12.14).
- Depending on the resolution of the staining procedure, additional bands may be detected within each region. These are designated by adding a digit to the number of the region (Figure 12.14).

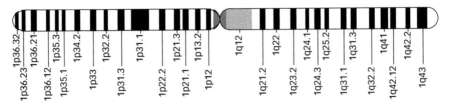

Figure 12.14 Ideogram of Giemsa-stained human chromosome 1.

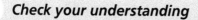

Check your understanding

12.7 Write the ISCN nomenclature for (i) a female with a Robertsonian translocation between q10 of chromosome 13 and q10 of chromosome 14; and (ii) a male with an inversion in the long arm of chromosome 13 between q14.3 and q21.2.

12.5 CHROMOSOMES AND CANCER

- All malignant tumour types have been found to contain cells with chromosome aberrations.
- Chromosome mutations are both:
  - the cause of cancer
  - the result of cancer.
- Various haematomological neoplasms (e.g. chronic myeloid leukaemia (CML)) and a few solid tumour types (e.g. Ewing's sarcoma) are associated with specific chromosomal abnormalities (Table 12.10).
- Furthermore, these specific chromosomal abnormalities correlate with specific gene alterations, e.g. a fusion gene product expressed constitutively (CML).
- More commonly, general genetic instability occurs in most advanced tumours resulting in cells with a range of non-specific chromosomal abnormalities of both structure and number.
- As a result, altered expression of oncogenes and tumour suppressor genes occurs.

Chronic myeloid leukaemia (CML)

- About 95% of individuals with CML have a reciprocal translocation between chromosomes 9 and 22 in their myeloid-derived white blood cells.
- The ISCN description of CML is t(9;22)(q34;q11).
- The chromosome 9 breakpoint is in a proto-oncogene termed *c-ABL*.

Chromosome mutation	Cancer	Description
t(9;22)(q34;q11)	Chronic myeloid leukaemia	Activates proto-oncogene c-ABL (a tyrosine kinase) on chromosome 9
t(8;14)(q24;q32)	Burkett's lymphoma	Deregulation of proto-oncogene c-myc (a mitotic transcription factor) at 8q24
t(11;14)(q13;q32)	Mantle cell lymphoma	Deregulation of cell cycle control gene cyclin D1
t(11;22)(q24;q11.2-12)	Ewing's sarcoma	Activation of *EWS* gene (encoding a RNA binding protein) on chromosome 22
inv 10 (q11.2;q21.2)t(10:17) (q11.2;q23)	Papillary thyroid carcinoma	Activates proto-oncogene RET (a receptor tyrosine kinase) on chromosome 10

Table 12.10 Specific chromosomal abnormalities and associated cancers

Chromosomes and cancer

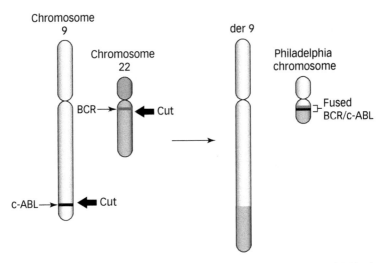

Figure 12.15 Formation of the Philadelphia chromosome, in chronic myeloid leukaemic individuals, via a reciprocal translocation.

- The break in chromosome 22 occurs in the *BCR* (breakpoint cluster) gene.
- The derived chromosome 22 is called the **Philadelphia chromosome** and has the 5′ section of *BCR* fused with most of *c-ABL* (Figure 12.15).
- Depending upon the *BCR* breakpoint a variable-sized fusion protein is produced.
- p190 fusion protein is the most common; p210 and p230 are rare variants.
- The hybrid *BCR–ABL* gene constitutively produces an abnormal 'fusion' tyrosine kinase, which activates various cell cycle control proteins. Increased unregulated mitosis results.
- The resulting cells are precancerous. A further mutation, e.g. in the tumour suppressor gene *p53*, transforms the cells into rapidly proliferating malignant cells.
- CML is treated with tyrosine kinase inhibitors, imatinib, and, more recently, dasatinib and nilotinib.

Looking for extra marks?

Most chromosomes have specific **fragile sites** where they are prone to break. About 50% of cancer-specific translocations have their break point in fragile sites. A few fragile sites (~5%) are inherited in a Mendelian fashion. For example, human fragile X syndrome (associated with mental retardation) is caused by a fragile site at Xq27.3. It preferentially affects males.

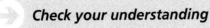 ### Check your understanding

12.8 Why do translocations or inversions in somatic cells sometimes lead to the development of cancer?

12.6 DETECTING CHROMOSOMAL MUTATIONS

- Chromosomes are examined microscopically following their treatment with:
  ○ chemical stains (**karyotyping**)
  ○ fluorescent DNA probes (**fluorescent *in situ* hybridization (FISH)**).
- Compared with karyotyping (see section 1.2), FISH is:
  ○ faster
  ○ more sensitive—small-scale chromosomal changes can be detected
  ○ is not restricted to cells in metaphase of mitosis.

Fluorescent *in situ* hybridization (FISH)

- A fluorescent antibody-tagged length of single-stranded DNA (the **probe**) is hybridized to metaphase or interphase chromosomes (Figure 12.16).
- Cells are mounted on microscopic slides under DNA denaturing conditions.
- The fluorescent probe is added and undergoes complementary base pairing (hybridization) with relevant DNA sequences.
- Depending on the target DNA there may be one or multiple complementary sequences.
- Probe hybridization is visualized under a fluorescent microscope.
- Multiple probes with different coloured dyes can be used simultaneously to investigate different sequences or chromosomes.
- FISH is informative within the contexts of genetic counselling, prenatal testing, and medical research.
- It is used:
  ○ to detect structural chromosome changes, particularly deletions—FISH produces much greater resolution than Giemsa banding and can detect deletions as small as 1 Mb (**microdeletions**)

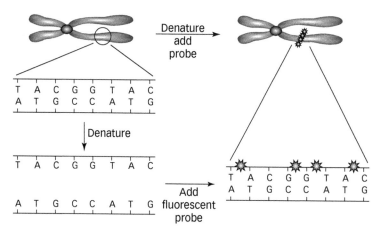

Figure 12.16 Principles of fluorescent in situ hybridization. The fluorescent DNA probes hybridize to complementary DNA sequence under denaturing conditions

Detecting chromosomal mutations

- to identify missing or extra chromosomes—FISH can be carried out with interphase chromosomes, so is quicker and cheaper than karyotyping in prenatal testing; furthermore, multiple probes enable testing for most common abnormalities in a single reaction
  - to study spatial and temporal patterns of gene expression as FISH can detect and localize specific mRNAs within tissue samples.
- Three types of FISH probes are used: centromeric, locus-specific, and chromosome-specific.
  - **Centromeric probes** bind to the unique alpha satellite sequences of centromeres. They give a diffuse, but intense, signal, and are useful to detect aneuploidies, e.g. human trisomies 13, 18, and 21.
  - **Locus-specific probes** give a sharp, discrete signal. They are particularly used to identify microdeletion syndromes, e.g. Williams or diGeorge syndromes.
  - **Chromosome-specific probes** are designed to colour the entire chromosome. By using FISH probes with different coloured fluorophores, each human chromosome can be identified independently. This is useful for detecting structural rearrangements (e.g. translocations, rings, isochromosomes) in tumour cells.

 Check your understanding

12.9   What techniques are available to a cytogeneticist to identify chromosomal abnormalities?

12.10 A cytogenetic screen reported a karyotype of 46,XX,dup(1)(p22.1 p31.1) in one patient and 46,XX,del(1)(p22.1 p31.1) in a second. How might the two chromosomal abnormalities have arisen, and which patient is expected to be more severely affected and why?

online resource centre   You'll find guidance on answering the questions posed in this chapter—plus additional multiple-choice questions—in the Online Resource Centre accompanying this revision guide. Go to **www.oxfordtextbooks.co.uk/orc/thrive/** or scan this image:

13 Genetics of Populations

Population genetics examines the genetic makeup of populations and how a population's genetic composition changes with time. It has particular relevance for evolutionary studies and biological conservation.

Key concepts

- Population genetics analyses the patterns of genetic variation shown by *groups* of individuals, i.e. by populations (Table 13.1).
- This contrasts with the main concern of Mendelian genetics (Chapter 2) and, to a large extent, of quantitative genetics (Chapter 5), which both focus on the genotype of *individuals* and the genotypes resulting from single matings.
- Population geneticists address such questions as:
  - How much genetic variation exists in a particular population?
  - What processes control the amount of genetic variation in a population?
  - How might this genetic variation change with time?
  - What processes produce genetic divergence between populations?
- Population genetics explores the evolutionary processes that shape a population's genetic variation, i.e. mating systems, migration, mutation, population size, and selective forces.

continued

Calculating allele and genotype frequencies

- The analysis of genetic diversity in populations of endangered species helps formulate conservation policy.
- The genetic variation within and between different populations is described in terms of *frequencies* of alleles and resulting genotypes.

Term	Definition
Population	A local group of a single species within which mating is actually or potentially occurring
Frequency	A proportion, thus ranging in numerical value from 0 to 1; e.g. if 30% of the alleles at a 2-allele locus (A/a) are the dominant A, then the frequency of A=0.3 and frequency of a=0.7
Polymorphism	The existence of two or more alleles at a locus: the frequency of the most common being 0.99 or less
Gene pool	The common set of genes and their alleles shared by a group of interbreeding sexually-reproducing individuals

Table 13.1 Key terms

13.1 CALCULATING ALLELE AND GENOTYPE FREQUENCIES

The Hardy–Weinberg principle

- The **Hardy–Weinberg principle** (or **law**), developed by Godfrey Hardy and Weilheim Weinberg during the early 1900s, lies at the core of population genetics.
- It enables us to:
  - calculate allele and genotype frequencies
  - analyse the effect on these frequencies, and so upon genetic variation in a population, of different mating systems and of the evolutionary processes of natural selection, mutation, migration, and genetic drift.
- Consider a gene with two alleles (A) and (a). There are three possible genotypes (AA, Aa and aa).
- Hardy and Weinberg deduced, from considering basic Mendelian genetics, that these genotypes will be present in a population in the following proportions:

$$p^2 + 2pq + q^2$$

where p^2 is the frequency of homozygous dominant genotype (AA), 2pq is the frequency of heterozygous genotype (Aa), q^2 is the frequency of homozygous recessive genotype (aa), p is the frequency of dominant allele (A), and q is the frequency of recessive allele (a).

- For a gene with two alleles A/a whose frequencies are p and q, the sum of these two frequencies in the population being analysed must equal 1:

$$p + q = 1$$

- Furthermore, the sum of the resulting genotype frequencies will also equal 1:

$$p^2 + 2pq + q^2 = 1$$

- These relationships exist among individuals at the time of sampling a study population.
- The same proportions will also hold in succeeding generations if no evolutionary forces are acting on a population to change allele frequencies.

> These ideas are formally stated in the Hardy–Weinberg principle: *allele and genotype frequencies in a population remain constant from generation to generation as long as the population is large, randomly mating and not affected by selection, migration or mutation.*

- When a population's genotypes are in the expected proportions of p^2, $2pq$, and q^2 the population is said to be in **Hardy–Weinberg equilibrium** (see section 13.2).
- Outside of the laboratory, where conditions can be controlled, one or more of the evolutionary processes are always in effect to some extent. For example, rarely are all genotypes of equal adaptive potential.
- Hardy–Weinberg equilibrium is impossible in nature. However, it is an ideal state that provides a baseline against which to detect and measure change.

Calculating genotype frequencies

- If the allele frequencies among individuals in a population are known, **genotype frequencies** can be calculated easily by substituting these values into the terms p^2, $2pq$, and q^2 (Box 13.1).

Calculating allele frequencies

- Allele frequencies can be calculated in two ways:
  - *directly* from the numbers of individuals expressing the different genotypes
  - *indirectly* from the frequency of the recessive genotype.
- The method used depends upon the dominance relationships of the alleles.
- When each genotype determines a separate phenotype (third column, Table 13.2) the individuals of each genotype can be identified and counted in a population. Allele frequencies can be determined directly (method 1).
- When one allele is completely dominant, AA and Aa give the same phenotype; an indirect method is used to determine allele frequencies (method 2).

(i) Method 1: Allele frequencies calculated directly from phenotype numbers

- This method is used when alleles are incompletely or co-dominant

$$\text{The frequency of each allele} = \frac{\text{number of copies of an allele}}{\text{total number of alleles}}$$

- This method is best explained by working through an example (Table 13.3).

Box 13.1 Calculating genotype and resulting phenotype frequencies

If the frequency in a population of allele $A=0.6$ and of allele $a=0.4$, then the frequency (f) of each genotype is calculated thus:

$$f(AA) = p^2 = (0.6)^2 = 0.36$$
$$f(Aa) = 2pq = (2 \times 0.6 \times 0.4) = 0.48$$
$$f(aa) = q^2 = (0.4)^2 = 0.16$$

These genotype frequencies enable us to predict phenotype frequencies. These will depend on the dominance relationships between alleles.

(i)   If allele A is completely dominant to allele a:

$$\text{Freq. dominant phenotype} = f(AA) + f(Aa)$$
$$= 0.36 + 0.48$$
$$= \mathbf{0.84}$$
$$\text{Freq. recessive phenotype} = f(aa)$$
$$= 0.16$$

(ii)   If alleles A and a are incompletely or co-dominant, the phenotype frequencies will be the same as the genotype frequencies calculated previously:

$$\text{Freq. dominant phenotype determined by } AA = 0.36$$
$$\text{Freq. dominant phenotype determined by } Aa = 0.48$$
$$\text{Freq. recessive phenotype} = 0.16$$

Genotypes	Allele A completely dominant to allele a	Allele A incompletely dominant to allele a
AA	Phenotype 1	Phenotype 1
Aa		Phenotype 2
aa	Phenotype 2	Phenotype 3

Table 13.2 Relationships between genotypes and resulting phenotypes (consider one gene with two alleles A and a)

(ii) Method 2: allele frequencies calculated indirectly from recessive phenotype frequency

- If one allele is completely dominant the numbers of different genotypes cannot be derived directly from phenotype numbers because the dominant phenotype includes both homozygotes (AA) and heterozygotes (Aa), which are indistinguishable phenotypically.
- However, the recessive phenotype is determined by a single genotype (aa) whose frequency is q^2.

Calculating allele and genotype frequencies

Phenotype	Number of individuals	Genotype	Number of alleles*	
			G	B
Green	280	GG	280×2=560	
Brown/green	130	BG	130	130
Brown	90	BB		90×2=180
Total	500*		690	310
	Frequency of allele G = $\dfrac{690}{1000^*}$ = 0.69		Frequency of allele B = $\dfrac{310}{1000}$ = 0.31	

Table 13.3 Calculating allele frequencies in a population of Australian banjo frogs. The two alleles B (brown) and G (green) are incompletely dominant

*Five hundred individuals share 1000 alleles as each individual possesses two copies of a gene.

- Thus:

Recessive **allele frequency** $(q)=\sqrt{q^2}$

Dominant allele frequency $(p)=1-q$ (derived from $p+q=1$)

- Box 13.2 illustrates the use of recessive phenotypes to calculate allele frequencies.
- Once allele frequencies are calculated, all genotype frequencies can then be determined.

Box 13.2 Calculating allele frequencies from recessive phenotype data

- Tongue-rolling ability is determined by a single autosomal gene with two alleles (T/t). The ability to roll your tongue is dominant (T).
- Nine hundred and fifty people were observed and 370 individuals could not roll their tongue.
- From these data the frequencies of the two alleles can be calculated.
- We know the number of individuals expressing the recessive phenotype (i.e. 370). From this observation, we can calculate the *frequency* of the recessive phenotype.

f(recessive phenotype frequency) = 370/950 = 0.39

thus

f(recessive genotype) = 0.39 = q^2

$q = \sqrt{q^2}$

$= \sqrt{0.39}$

$= 0.624$ (*frequency of t allele*)

$p = 1-q$

$= 1-0.624$

$= 0.376$ (*frequency of T allele*)

Calculating allele and genotype frequencies

- Continuing the example of Box 13.2, the frequency of individuals who are homozygous (TT) and heterozygous dominant (Tt) can be determined (as can the absolute *number* of individuals):

$$f(TT) = p^2 = (0.376)^2 = 0.141$$
$$f(Tt) = 2pq = 2 \times 0.376 \times 0.624 = 0.47$$

- In our population of 950 individuals we would expect:

Number of TT individuals = 0.141 × 950 = 133.95 (134 to nearest whole number)

Number of Tt individuals = 0.47 × 950 = 445.8 (so 446)

This calculation illustrates an important cautionary point: always be clear whether you are dealing with **frequencies** *or* **numbers**.

- This example also illustrates another important use of population genetics in the context of **medical genetics**. It enables calculation of population carrier (heterozygous) numbers for severe inherited recessive conditions (Box 13.3).
- Note that when the frequency of an allele is low most copies of the recessive allele are present in heterozygotes (Box 13.3).

Box 13.3 Calculating carrier frequencies for human inherited recessive conditions

- Wilson's disease (a failure to metabolize copper) is caused by a single autosomal recessive allele.
- Its incidence is 1 in 33,000.
- From this data the carrier (heterozygote) frequency can be calculated.

f(recessive phenotype frequency) = f(recessive genotype) = q^2

thus

$$q^2 = 1 \text{ in } 33,000 \ (3 \times 10^{-5})$$
$$q = \sqrt{q^2}$$
$$= \sqrt{(3 \times 10^{-5})}$$
$$= 0.0055 \text{ (frequency of Wilson's disease allele)}$$
$$p = 1 - q$$
$$= 1 - 0.0055$$
$$= 0.9945$$
$$f(\text{heterozygote}) = 2pq = 2 \times 0.9945 \times 0.0055$$
$$= 0.011$$

- A carrier frequency of 0.011 means that 1 in 91 individuals is a carrier of the recessive Wilson's disease allele.

Calculating allele and genotype frequencies

	Gene has three alleles: A^1, A^2, A^3	X-linked gene with two alleles: H, h
Allele frequencies	$f(A^1)=p$ $f(A^2)=q$ $f(A^3)=r$	$f(H)=p \quad f(h)=q$
Genotype frequencies	$f(A^1 A^1)=p^2$ $f(A^1 A^2)=2pq$ $f(A^2 A^2)=q^2$ $f(A^3 A^3)=r^2$ $f(A^1A^3)=2pr$ $f(A^2A^3)=2qr$	Females: $f(X^HX^H)=p^2$ $f(X^HX^h)=2pq$ $f(X^hX^h)=q^2$ Males: $f(X^HY)=p$ $F(X^hY)=q$

Table 13.4 Allele and genotype frequencies for X-linked and multiple allelic genes

Calculating allele and genotype frequencies for sex-linked and multiple allelic genes

- The same Hardy–Weinberg principles apply.
  - The sum of all allele frequencies$=1$.
  - The expected genotype frequencies can be calculated from allele frequencies.
- Consider a gene with three alleles (Table 13.4):
  - allele frequencies are represented by **p**, **q**, and **r**
  - thus, **p+q+r=1**
  - the different genotype frequencies are $p^2+2pq+q^2+r^2+2pr+2qr$.
- Consider an X-linked gene with two alleles:
  - the frequencies of these two alleles are represented by p and q, thus $p+q=1$
  - male and female genotype frequencies are different as a female has two X chromosomes (and therefore two copies of an allele) and males have one X chromosome (and one copy of each X-linked allele) (Table 13.4).

Check your understanding

13.1 Full (dark) colour in cats is dominant to dilute (pale) colouring. Thirty-six per cent of domestic cats in a city had the homozygous recessive genotype (**dd**). Calculate: (i) the frequency of the **d** allele; (ii) the frequency of the **D** allele; (iii) the frequencies of the two possible phenotypes.

13.2 Within a population of butterflies, brown (B) is dominant over white (b). Twenty-four per cent of butterflies are white. Calculate: (i) the percentage of butterflies in the population that are heterozygous; (ii) the proportion of brown individuals that are homozygous.

13.3 The prevalence of haemophilia A (X-linked recessive) among the females of a population was found to be 1/1,000,000. (i) What is the recessive allele frequency in this population? (ii) What would be the incidence of haemophilia among males?

13.2 HARDY–WEINBERG EQUILIBRIUM

- The Hardy–Weinberg principle states that, for the locus under consideration, *if* a population is large, is randomly mating and is not affected by selection, mutation, and migration:
  - allele frequencies of a population are constant in succeeding generations
  - these allele frequencies determine genotype frequencies in the predictable proportions of p^2, $2pq$, and q^2.
- Thus, if a population is sampled and genotypes are in their expected proportions of p^2, $2pq$, and q^2, the population is said to be in Hardy–Weinberg equilibrium.
- At equilibrium, evolutionary forces, such as selection, mutation, and migration, are not operating on the population, with respect to the locus under consideration.

Testing for Hardy–Weinberg equilibrium

- To gain insight into whether or not there are evolutionary forces of change acting on a population, a test for Hardy–Weinberg equilibrium is performed.
- The *expected* numbers of individuals of each genotype, if the population is at equilibrium for that locus, are compared with the *observed* numbers of individuals of each genotype using a **goodness-of-fit chi-square (χ^2) test**.
- If there is no significant difference between observed and expected proportions Hardy–Weinberg equilibrium exists.
- For a χ^2 test to be possible there has to be a means of directly determining a population's genotype frequencies to produce observed frequencies, e.g. genotypes produced by incompletely dominant alleles (section 13.1) or identifiable by molecular analysis (Box 13.4).
- Expected genotype frequencies (p^2, $2pq$, and q^2) are determined from the allele frequencies calculated for the population under study.

Disturbances of Hardy–Weinberg equilibrium

- Hardy–Weinberg equilibrium is disrupted if:
  - mating is non-random
  - mutation, migration, genetic drift, or selection of certain genotypes occurs.
- These processes variously alter allele and genotype frequencies among individuals in a population (Table 13.5).
- In the following sections each factor is considered separately, but several factors will be interacting simultaneously in a natural population. For example, non-random mating preferentially increases certain genotypes which might be at a selective disadvantage.

Box 13.4 Testing for equilibrium

Occasionally, white bears appear in populations of the black bear *Ursus americanus* along the coast of British Columbia. White fur results from a recessive mutation in the *MC1R* pigmentation gene. Genotypes (AA/AG/GG) formed from the two alleles (A: black; G: white) were identified by sequencing DNA extracted from 87 bears from 3 islands. A χ^2 test was used to assess whether this population was in Hardy–Weinberg equilibrium.

Firstly, **observed** allele frequencies were calculated.

Number of bears	Genotype	Number of alleles	
		A	G
Black: 42	AA	84	
Black: 24	AG	24	24
White: 21	GG		42
Total: 87		108	66
Frequency of allele A $= \dfrac{108}{174} = 0.62$		**Frequency of allele G** $= \dfrac{66}{174} = 0.38$	

These observed allele frequencies were then used to calculate **expected** genotype frequencies and numbers in the island population using the Hardy–Weinberg frequencies of p^2, $2pq$, and q^2

Genotype	Expected frequency	Expected numbers
AA	0.384	33.4
AG	0.471	41
GG	0.144	12.6

A χ^2 test is used to assess whether there is any significant difference between observed and expected genotype proportions.

Null hypothesis: there is no difference between observed and expected numbers of bears of genotypes AA/AG/GG.

	Observed numbers (O)	Expected numbers (E)	O – E	(O – E)²	$\dfrac{(O-E)^2}{E}$
AA	42	33.4	8.2	67.24	2.01
AG	24	41	−17	289	7.05
GG	21	12.6	8.4	70.56	5.6
Total	87	87			**14.66**

Significance level $= 0.05$; 2 degrees of freedom, calculated χ^2 value $= 14.66$; critical χ^2 value $= 5.99$.

As the calculated χ^2 value is greater than the critical χ^2 value the null hypothesis is rejected. This indicates that there is a significant difference between observed and expected genotype frequencies. The population of black bears is *not* in Hardy–Weinberg equilibrium with respect to coat colour.

	Allele frequencies	Genotype frequencies
Non-random mating	Unchanged	Changed
Mutation, migration, drift, and selection	Changed	Changed

Table 13.5 The effect of evolutionary processes on allele and genotype frequencies

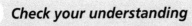

Check your understanding

13.4 Under what conditions do allele frequencies stay constant from one generation to the next?

13.5 How can you determine whether a population is in Hardy–Weinberg equilibrium?

13.3 NON-RANDOM MATING

- There are two types of non-random mating:
  - inbreeding/outbreeding—affects *all* gene loci simultaneously
  - **assortative mating**—affects *only* the loci (and linked genes) involved in expression of a phenotypic trait used in mate choice.
- **Inbreeding** involves preferential mating between close relatives. It is also referred to as **consanguineous mating**.
- **Outbreeding** is the avoidance of mating between related individuals, so mating individuals are less related than expected if they had been drawn by chance from a random mating population.
- Assortative mating involves either:
  - mating between individuals that share a particular phenotype (**positive assortative mating**), e.g., tall people preferentially mate with other tall people
  - avoidance of mating between individuals that share a particular phenotype (**negative assortative mating**), e.g. human immune systems.
- **Disassortative mating** refers to preferential mating between phenotypically dissimilar individuals.

A closer look at inbreeding

- Inbreeding changes a population's genotype frequencies, but has no effect on allele frequencies.
- It leads to:
  - an increase in the proportions of homozygotes (AA and aa)
  - a decrease in the proportion of heterozygotes (Aa).
- To emphasize the loss of heterozygosity associated with inbreeding, consider self-fertilization—the most extreme case of inbreeding (Table 13.6).

Generation	Genotype frequencies (%)		
	AA	Aa	aa
F_1	0.25	0.5	0.25
F_2	0.375	0.25	0.375
F_3	0.437	0.125	0.437
F_4	0.468	0.063	0.468
F_n	$\dfrac{1-1/2^n}{2}$	$1/2^n$	$\dfrac{1-1/2^n}{2}$

Table 13.6 Reduction in heterozygote frequency by self-fertilizing individuals (parental self-fertilizing individual was Aa)

- The problems associated with inbreeding arise from the increase in recessive homozygotes. Their expression is often deleterious or lethal.
- Most populations carry deleterious recessive alleles, but these are 'hidden' in the heterozygotes of a randomly mating population in Hardy–Weinberg equilibrium.
- For example, if a recessive allele (q) has a frequency of 0.01, 99% of the alleles are present in heterozygotes ($2pq=0.0198$; $q^2=0.001$).
- The reduction in fitness in inbred populations is termed **inbreeding depression**, which involves:
  - reduced fecundity (lowered sperm viability, smaller litter size, higher infant mortality)
  - a slower growth rate and smaller adult size
  - reduced immune system function.
- Inbreeding depression is only a problem when mating occurs between related individuals in a previously randomly mating population.
- Some species habitually inbreed. For example, various plant species self-fertilize and sibling mating occurs in some animal species, such as gif wasps and certain parasites.
- In those species in which inbreeding is the norm, deleterious genes are likely to have been purged from the population such that the population is homozygous for beneficial characters.
- In the evolutionary short term, an inbred population may have an advantage: more individuals possess the alleles, and so possess those genotypes and phenotypes to adapt them to prevailing environmental conditions. In the long term, inbred populations lack evolutionary flexibility.

The inbreeding coefficient

- The extent of inbreeding in a population is quantified by the **inbreeding coefficient (F)**.
- The inbreeding coefficient is defined as the probability that two alleles at a locus are identical by descent from a common ancestral allele.
- F has a value between 0 and 1.

Non-random mating

- A value of F=0 indicates that mating in a population is random and also that the population is likely to be large.
- The more inbred a population, the higher the value of F.
- F can be calculated in two ways:
  - from analysis of pedigrees
  - by considering the reduction in observed heterozygosity in a population, relative to what you would expect from observed allele frequencies.

(i) Calculating F by pedigree analysis

- To use this method to calculate F you need to be able to construct pedigrees and identify pathways of inheritance of ancestral alleles.
- In a homozygous individual a distinction can be made as to whether the two alleles are:
  - **identical by descent (IBD)**
  - **identical by state (IBS)**.
- Consider the origins of the two copies of allele *a* in the two homozygous individuals in each pedigree in Figure 13.1.
  - In pedigree 13.1A the two copies of allele *a* are descended from the same copy of allele *a* in a common ancestor (the great grandmother of IV-1). They are **IBD**.
  - In pedigree 13.1B the two copies of allele *a* are descended from two different copies of allele *a*. They are **IBS**.
- It is IBD-type pedigree that must be used for calculating F.
- An example of using an IBD pedigree to calculate a value for F is given in Box 13.5.
- There are only a few situations when it is possible to obtain sufficient data to be able to construct pedigrees that can be used to calculate F.
- These rare opportunities include human genetic counselling and captive breeding programmes in zoos.
- More often, a natural population is sampled at one moment in time with no knowledge of past matings.
- In such situations F is calculated from an analysis of population heterozygosity levels.

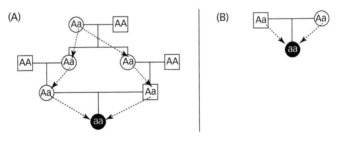

Figure 13.1 Alleles of homozygotes: identical by descent (IBD) or by state (IBS). (A) Alleles *a* in homozygote IV-1 are IBD: III-1 and III-2 share same maternal grandmother. (B) Alleles *a* in homozygote II-1 are IBS: I-1 and I-2 are unrelated.

Box 13.5 Calculating the inbreeding coefficient (F) from pedigree analysis

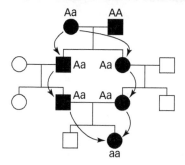

1. Consider individual IV-2 (the child of a first cousin marriage).
2. Identify the two paths that alleles must take to become **IBD** (indicated by arrows).
3. For each path calculate the number of generations from the common ancestor to IV-2 (three generations for each path).
4. Calculate the probability of **IBD** for each allele for each path.
5. The probability that an allele is **IBD** between two generations is **1/2**. In each case you are considering the possibility that a heterozygote (Aa) passes on its recessive allele (a) to the relevant individual of the next generation, e.g. II-2 to III-3.
6. Thus, the probability of **IBD** for each path for each allele is $(1/2)^3$.
7. Multiply the probabilities of each path—$(1/2)^3 \times (1/2)^3$—to obtain the total probability $((1/2)^6$ or 1/64) of **IBD** for that allele.
8. Thus, for great-grandmother (I-1)'s allele **a** the chance that individual IV-2 inherits a copy of this allele is 1/64.
9. *To obtain a value for F you need to consider that IV's two alleles could be IBD from any one of the four alleles of her great grandparents.*
10. Thus, $F = 4 \times 1/64 = 1/16$ (or 0.0625).

(ii) Calculating F from population heterozygosity levels

- In an inbred population the observed number of heterozygotes will be less than the expected number from Hardy–Weinberg predictions, i.e. from the number calculated from observed allele frequencies.

- This is because alleles are preferentially homozygotes compared with heterozygotes in inbred populations.

- Thus, a comparison of expected-to-observed number of heterozygotes gives a measure of inbreeding, i.e. a value for the inbreeding coefficient, F.

- If H = observed heterozygosity and 2pq = expected heterozygosity:

$$F = \frac{2pq - H}{2pq}$$

Non-random mating

- F is frequently partitioned into F_{IS} and F_{ST}.
  - F_{IS} measures whether there are fewer or more heterozygous individuals than expected in a subpopulation.
  - F_{ST} (or θ) measures degree of genetic differentiation among subpopulations.

Looking for extra marks?

F_{ST} is much used in conservation genetics to indicate how distinct fragmented populations are. Values for F_{ST} are frequently produced from molecular data, such as microsatellite alleles.

Calculating genotype frequencies in inbred populations

- Table 13.6 showed that for each generation of inbreeding the increase in the proportion of each homozygote is equal to *half* of the proportionate decrease in heterozygotes.
- The proportion of heterozygotes being removed and appearing as homozygotes relates to the level of inbreeding and so to the value of F.
- If you know the value of F and allele frequencies, the expected genotype frequencies in the inbred population will be:

$$f(AA) = p^2 + Fpq$$
$$f(Aa) = 2pq - 2Fpq$$
$$f(aa) = q^2 + Fpq$$

- NB, inbreeding reduces heterozygosity—and, hence, future genetic variability—in a population.
- To emphasize the deleterious effect of inbreeding in a normal, randomly mating population consider the consequences of first cousins marrying (Box 13.6).

Box 13.6 *Calculating the frequency of the recessive homozygote in an inbred population*

- Assume that a recessive allele that causes an inherited disease has a population frequency (q) of 0.1.
- If there is random mating, F=0 and $q^2 = 0.01^2 = 0.0001$ (1 in 10,000 individuals are homozygous and express the condition).
- If first cousins marry, F=0.0625.
- Expected frequency of homozygotes $= q^2 + Fpq = 0.01^2 + (0.0625)(0.99)(0.01) = 0.00062$ (6.2 in 10,000 individuals are homozygous and express the condition).
- Expression of this inherited disease is 6.2 times more likely.
- For siblings (F=0.25) the increased likelihood rises to 26 times.

Check your understanding

13.6 What are the causes and effects of inbreeding depression?

13.7 What is inbreeding? How is it measured?

13.4 MUTATION

Gene mutations (Chapter 11) are an important evolutionary force as they are the *only* means of producing new alleles and so *new* variation.

- Mutations are not, however, a major force in changing allele frequencies as:
  - their rates are too low
  - a single gene mutation results in an extremely small change in allele frequencies (Box 13.7).

Box 13.7 Modelling the effect of mutation on allele frequencies

A gene has two alleles (**A/a**) with frequencies $p=0.9$ and $q=0.1$.

Forward mutation rate μ $(A \rightarrow a)=1 \times 10^{-5}$

Reverse mutation rate v $(a \rightarrow A)=0.3 \times 10^{-5}$

Changes in allele frequencies, as the result of mutation, are determined by:

1. Mutation rate
2. Existing allele frequencies.

   Consider frequencies of alleles **A** and **a** in the generation following the mutation.

   Change in frequency of $A = \mu \times p$ (μp)

   Change in frequency of $a = v \times q$ (vq)

   To determine the **net change** in allele frequencies per generation you need to consider both forward and reverse mutational changes. For example, net change in frequency of allele **a** is represented by Δq.

   $$\Delta q = \mu p - v$$

Substituting values:

$$\Delta q = (1 \times 10^{-5} \times 0.9) - (0.3 \times 10^{-5} \times 0.1)$$
$$= 8.7 \times 10^{-6}$$

Thus, the frequency of allele **a** has changed from 0.1 to 0.1000087 in the first generation.

Migration

- A significant increase in the frequency of a new allele will only occur in combination with other evolutionary processes, such as migration or selection.
- Mutation frequencies are generally expressed as the number of new alleles per locus per given number of gametes: typically, for eukaryotes, 1 new mutation per locus per 100,000 gametes (i.e. 10^{-5}) per generation.
- For ease of modelling, the mutational effects considered in Box 13.7 were forward and reverse events between the two existing alleles of a gene.
- In reality, a mutation generally produces a new allele. Initially, this new allele will be at an extremely low frequency.
- Furthermore, a recessive mutation is likely to be present for many generations in a heterozygote where its adaptive evolutionary potential is untested, i.e. an allele is not available to be acted upon by selection until present in a homozygote when its phenotypic consequence is expressed.
- A mutation may be **neutral**, beneficial, or harmful. A beneficial mutation is rare. The majority of mutations will be neutral or harmful.
- It is only over the course of many generations (often thousands) that a mutation will result in a significant change in allele, and so genotype, frequencies.

13.5 MIGRATION

- Migration is the movement of individuals from one geographically separate population to another.
- Migration moves alleles between populations and thus changes allele frequencies.
- It may be a one-off event or a continual process, when it is also termed **gene flow**.
- Migration:
  - increases genetic variation within a population by introducing new alleles
  - decreases genetic differences between populations.
- From a conservation perspective, migration of individuals into small populations counters the detrimental effects of drift and inbreeding, and so preserves genetic diversity.
- However, in abolishing genetic differentiation between populations, migration counters evolutionary processes towards speciation, as it reduces local adaptation.
- The magnitude of allele frequency change achieved by migration depends on:
  - the extent of migration (how many individuals move from the source population)
  - the difference in allele frequencies between source and recipient populations.
- The effect of migration on allele frequencies can be calculated easily (Box 13.8).
- Box 13.8 shows how, if allele frequencies in migrating and recipient populations are known, migration rates can be determined. This has particular relevance in the context of the important conservation issue of **introgression**, when the genetic integrity of a species is threatened by migration into a population of individuals from another unrelated population or subspecies.

Box 13.8 Changes in allele frequencies due to migration

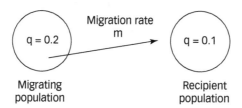

Migrating
population

Recipient
population

- q_m = recessive allele frequency in the source (migrating) population.
- q_o = recessive allele frequency in the recipient (original) population.
- m = proportion of migrating individuals.

The new recessive allele frequency after migration is calculated using the following equation:

$$q_n = q_m(m) + q_o(1 - m)$$

So if 20% of the population migrated and $q_m = 0.2$ and $q_o = 0.1$, then the recessive allele frequency in the next generation would be 0.12.

If migration continues the difference in allele frequency declines exponentially. The following graph illustrates the consequences of five populations exchanging migrants at the rate of $m = 0.1$ per generation.

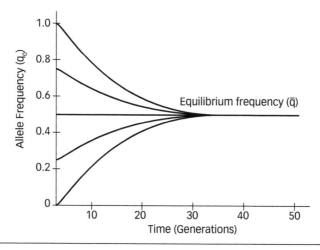

Check your understanding

13.8 Compare and contrast the impact on allele frequencies made by mutation and migration. Which is the stronger evolutionary force?

13.6 SELECTION

- Natural selection acts on phenotypes, resulting in the preferential survival of individuals with certain phenotypes (and, hence, a differential likelihood of the underlying genotype being passed on).
- Selection is the agent that brings about adaptive evolution. It can be a powerful tool for change. For example, in the case of the peppered moth, *Biston betularia,* the predominant form changed from pale to dark in selective response to the darkening of its background environment during the Industrial Revolution of the nineteenth century.
- Selection can be (Figure 13.2):
  - **directional**—favouring either homozygous dominant or homozygous recessive individuals
  - **balanced**—when positive and negative selection forces counter each other, establishing an equilibrium with constant allele frequencies; this occurs with **heterozygous advantage** when the heterozygote is favoured relative to the two homozygotes, e.g. heterozygotes for the sickle-cell anaemia allele have the highest fitness as they do not suffer from anaemia and are malaria resistant
  - **disruptive**—when heterozygotes are at a selective disadvantage; instead, there is positive selection of both dominant and recessive homozygotes.
- As selection favours certain genotypes, allele frequencies change in succeeding generations.
- Various formulae have been developed to calculate new allele frequencies, depending on:
  - type of selection
  - intensity of selection.
- Selection equations:
  - all derive from the two basic Hardy–Weinberg equations ($p+q=1$; $p^2+2pq+q^2=1$)

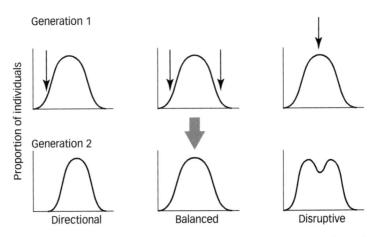

Figure 13.2 The three types of selection: (A) directional; (B) balanced; (C) disruptive.

- incorporate one of the two coefficients:
  - the **fitness coefficient (w)**, a measure of the relative reproductive success of different genotypes formed by a set of alleles
  - the **selection coefficient (s)**, a measure of the relative intensity of selection against different genotypes.
- Values for both coefficients range from 0 to 1 (Table 13.7):
  - the fitter a genotype, the higher the value of **w**
  - a higher **s** value indicates strong selection against a genotype.
- The two coefficients are related in the following way:

  $s=1-w$

  $w=1-s$

- The coefficients are calculated from survival data of individuals of different genotypes (Table 13.7).
- The fitness coefficients for different genotypes in a population can be used to calculate genotype and allele frequencies in succeeding generations (Table 13.8).
- NB, genotype frequencies in a population should equal 1 ($p^2+2pq+q^2=1$). Selection removes individuals from the population. Thus, the sum of a population's genotype frequencies immediately following selection is less than 1 (Table 13.8, row 3).
- The new summed value of genotype frequencies is designated $\overline{w}$. ($w=p^2wAA+2pqwAa+q^2waa$; here 0.664).
- $\overline{w}$ is used to **normalize** genotype frequencies, i.e. to re-establish the '$p^2+2pq+q^2=1$' relationship as shown in Table 13.8 (row 4).
- Table 13.9 presents three useful equations that can be used to calculate changed allele and, hence, genotype frequencies as the result of different types of selection.
- Practice is the key to understanding when and how to use the different equations. Many computer programs are now available to model different selection situations.
- Remember how useful mathematical evolutionary theory can be, e.g. it can give information about how rapidly antibiotic or pest resistance may arise in a species.

	A¹A¹	A¹A²	A²A²
Number of offspring surviving to adulthood	300	180	30
Fitness coefficient (w) *	$\frac{300}{300}=1$	$\frac{180}{300}=0.6$	$\frac{30}{300}=0.1$
Selection coefficient (s) (s=1−w)	0	0.4	0.9

Table 13.7 The relative fitness of three genotypes: AA, Aa, aa. Values for w are calculated from the numbers surviving to adulthood from every 500 individuals born
*To calculate fitness for each genotype we divide the number of surviving offspring with a certain genotype by the number of individuals produced by the most successful genotype.

Selection

Genotypes	AA	Aa	aa
Zygote (initial) genotype frequencies	p^2 (0.36)	$2pq$ (0.48)	q^2 (0.16)
Relative fitnesses (using values for w from Table 13.7)	wAA* (1.0)	wAa (0.6)	waa (0.1)
Relative genotype frequencies after selection	p^2wAA (0.36)	2pqwAa (0.29)	q^2waa (0.016)
Normalized genotype frequencies	$\dfrac{p^2wAA}{\overline{w}} = 0.54$	$\dfrac{2pqwAa}{\overline{w}} = 0.43$	$\dfrac{q^2waa}{\overline{w}} = 0.02$
Allele frequencies after selection	$p = \dfrac{p^2wAA + pqwAa}{\overline{w}}$	$q = \dfrac{q^2waa + pqwAa}{\overline{w}}$	

Table 13.8 Changes in genotype and allele frequencies produced by selection. Initially, p=0.6 and q=0.4

*Textbooks often present genotypes as A¹A¹, A¹A², and A²A², and so relative fitnesses as w_{11}, w_{12}, and w_{22}.

Selection type	Fitness values			Change in q (q)
	AA	Aa	aa	
Against a recessive trait	1	1	$1-s$	$\dfrac{-spq^2}{1-sq^2}$
Against a dominant trait	$1-s$	$1-s$	1	$\dfrac{-spq^2}{1-s+sq^2}$
Against both homozygotes (heterozygous advantage)	$1-sAA$	1	$1-saa$	$\dfrac{pq(sAA-saaq)}{1-sAAp^2-saaq^2}$

Table 13.9 Formulae for calculating allele frequencies following different types of selection

Mutation–selection balance

- There is an evolutionary balance between mutation and selection:
  - mutation produces new copies of a deleterious recessive allele
  - selection will eliminate these alleles, when their effect on fitness can be tested as a homozygote
- An equilibrium is reached in which the number of new alleles produced by mutation is balanced by the elimination of alleles through natural selection.
- The equilibrium frequency of q (**qe**) is calculated from the following equation:

$$q_e = \sqrt{u/s}$$

Check your understanding

13.9 Why is the elimination of a deleterious recessive allele by natural selection more difficult in a large than a small population?

13.7 GENETIC DRIFT

- Genetic drift is the random change in allele frequency in succeeding generations due to unequal sampling of gamete types during fertilization.
- All populations show a random variation in allele frequencies from generation to generation.
- In small populations, over many generations, genetic drift can have a major effect on allele frequencies.
- The effects of genetic drift on allele frequencies is frequently illustrated by Monte Carlo computer simulations (Figure 13.3).
- Monte Carlo simulations use a random number generator to choose an outcome for each probabilistic event; here, which gamete, with which allele, is used in fertilization.
- The simulation of Figure 13.3 emphasizes two key aspects of genetic drift.
  - The random nature of genetic drift: allele frequency is as likely to increase as decrease and 'wanders' with the passage of time, hence the term 'drift'.
  - The loss of genetic variation resulting from genetic drift. Through random change in succeeding generations an allele may be lost ($g=0$) or **fixed** ($g=1$).
- **Fixation** refers to the point in the process of genetic drift when the frequency of an allele$=1$. All other alleles of a gene have been lost from the population and all individuals will be homozygous for the fixed allele.

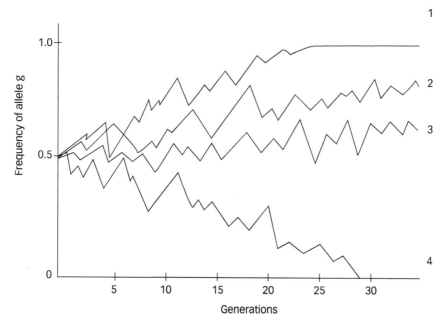

Figure 13.3 Genetic drift and changes in allele frequencies. The frequency of allele g is considered in 4 populations over 30 generations. Each population was small, consisting of 20 individuals; starting frequency of g$=0.5$.

Genetic drift

- Thus, genetic drift causes:
  1. Changes in allele frequencies
  2. Loss of genetic variation (through fixation of alleles)
  3. Genetic divergence between populations (the consequence of factors 1 and 2).
- These negative effects of genetic drift are strongly expressed in small populations.
- Small populations are produced in a variety of ways:
  - A large population may be **fragmented** into a number of small, non-communicating populations, e.g. through habitat destruction
  - a small number of individuals may establish a new population—the **founder effect**
  - a large population undergoes a sudden and drastic reduction in size—a genetic **bottleneck** occurs.
- In all three examples the initially small populations may increase in size. However, genetic variation will remain low because the alleles of each gene will be derived from those individuals who founded the new population or who survived the bottleneck.
- For example, a genetic bottleneck was produced in the northern elephant seal (*Mirounga angustisostris*) during the late nineteenth century through intensive hunting. Numbers have now recovered, but genetic variation remains very low in this species.
- The issues of genetic drift reducing genetic variation in small populations is of particular concern to conservationists, who incorporate strategies in their programmes to maximize allele variability. For example:
  - translocating individuals from larger populations to small ones with impoverished allelic diversity
  - controlling matings in captive breeding programmes
  - analysing and choosing individuals with maximum allele diversity for re-introduction programmes.

Looking for extra marks?

A striking human example of a population bottleneck is the expression of achromatopsia (total insensitivity to light due to lack of cones) among inhabitants of the Pacific atoll of Pingelap. In 1775 a typhoon reduced the population from 1000 to 20. Among the survivors, one male carried the recessive mutation for achromatopsia. Now about 1 in 12 of the 3000 Pingelap residents express achromatopsia compared with the normal 1 in 50,000–100,000.

Check your understanding

13.10 A geneticist is looking at 3 loci each with 2 alleles in a population of 40 plants of the endangered wood poppy (*Stylophorum diphyllum*). Allele frequencies are: A=0.68/a=0.32; B=0.84/b=0.16; and C=0.99/c=0.01. What are the relative probabilities that all three recessive alleles will be lost from this population through genetic drift?

13.11 How are the effects of genetic drift and inbreeding similar and different?

13.12 How does population genetics provide evidence to support the theory of evolution?

online resource centre You'll find guidance on answering the questions posed in this chapter—plus additional multiple-choice questions—in the Online Resource Centre accompanying this revision guide. Go to **www.oxfordtextbooks.co.uk/orc/thrive/** or scan this image:

14 Working with Genes: Analysing and Manipulating DNA

A wide range of molecular biology techniques enable DNA to be manipulated and analysed, yielding information about the nature and function of genes.

Key concepts

- Multiple copies of a DNA sequence can be produced by cloning or by using the polymerase chain reaction.
- Genes are isolated from DNA libraries.
- Gel electrophoresis separates different-sized DNA fragments.
- The nucleotide sequence of a segment of DNA is determined by Sanger's dideoxy method or next generation sequencing methods.
- Forward and reverse genetics are different analytical approaches to linking phenotype and genotype.

14.1 CLONING DNA

- The terms **recombinant DNA technology**, **DNA cloning**, and **gene cloning** all refer to the same process, namely the transfer of a DNA fragment from one organism to a self-replicating genetic element that replicates the fragment in a foreign host cell.

- It involves the following key stages:

  - DNA containing the gene or specific DNA sequence of interest is purified and chopped into fragments by **restriction enzymes**
  - DNA fragments are inserted into a self-replicating DNA molecule or **vector**
  - the vector is introduced into a new host cell where it is either maintained as a self-replicating unit in the cytoplasm or enters the host genome; the vector and/or host cell genome is manipulated so that recombinant cells can be identified.

Restriction enzymes

- Restriction enzymes are **endonucleases** that recognize specific short (4–6 base pairs) DNA sequences and make double-strand cuts.
- Sequences recognized and cut by restriction enzymes are termed **restriction sites**.
- DNA sequences recognized by restriction enzymes are palindromic.
- ➔ *See section 8.2 (p. 152) for details of palindromic sequences.*
- There are four types of restriction enzymes defined by:
  - the nature of their target sequence
  - the site of cutting relative to their recognition sequence
  - co-factor requirements.
- All restriction enzymes used in molecular genetics work are **type II**, i.e. they cut the DNA within, or close to, the recognition sequence.
- Some restriction enzymes stagger their cuts leaving single-stranded overhangs, called **sticky ends**.
- Other enzymes cut flush in the middle of their recognition sequence, producing **blunt ends** (Table 14.1).
- An enzyme that recognizes a 6-base pair sequence cuts, on average, every 4096 (4^6) base pairs, whereas one that recognizes a 4-base pair sequence cuts every 256 (4^4) base pairs.
- Restriction enzymes are specific: a particular enzyme originates from one bacterial species and recognizes a specific sequence (Table 14.1).
- In bacteria, restriction enzymes have evolved to recognize and digest invading phage DNA.
- Restriction sequences in bacterial genomes are methylated to protect them from restriction enzyme digestion.
- Typically, when working with restriction enzymes, a concentrated solution of purified DNA is digested in a small volume (25–50 µl) of buffer at 37°C for 2–3 hours. This produces a set of DNA or **restriction fragments**.

Chapter 14 Working with Genes: Analysing and Manipulating DNA   275

Cloning DNA

Enzyme*	Source	Recognition sequence	Cut	
EcoRI	*Escherichia coli*	5' GAATTC 3' 3' CTTAAG 5'	5' G 3' CTTAA	AATTC 3' G 5'
HindIII	*Haemophilus influenzae*	5' AAGCTT 3' 3' TTCGAA 5'	5' A 3' TTCGA	AGCTT 3' A 5'
HaeIII	*Haemophilus aegyptius*	5' GGCC 3' 3' CCGG 5'	5' GG 3' CC	CC 3' GG 5'

Table 14.1 Examples of restriction enzymes

*The enzyme naming system is based on bacterial genus, species, strain, and order of identification.

Vectors

- A cloning vector is a small, replicating DNA molecule into which a foreign DNA fragment can be inserted and transferred to a host cell.
- An effective cloning vector (Figure 14.1) needs three key characteristics:
  - (i) An **origin of replication**, so that it is replicated in the host cell
  - (ii) **Selectable markers**, e.g. antibiotic resistance, so that host cells containing the vector can be identified
  - (iii) A single restriction site for the enzyme used to generate the DNA fragment to be cloned, so that it can be inserted into the vector.
- Cloning vectors frequently have a **multiple cloning site** (MCS), a region which contains many restriction sites, enabling cloning of DNA generated by cleavage with many different restriction enzymes (Figure 14.1).
- There are four main types of cloning vectors: plasmids, phages/viruses, cosmids, and artificial chromosomes (Table 14.2). The choice of vector is decided by:
  - the size of the DNA fragment to be cloned
  - whether the DNA is being cloned in prokaryotes or eukaryotes
  - whether the cloned DNA is to be expressed.
- An **expression vector** contains additional DNA sequences, such as a promoter, transcription initiation and termination sites, and a Shine–Dalgarno sequence, so that the inserted DNA can be transcribed and, subsequently, translated.

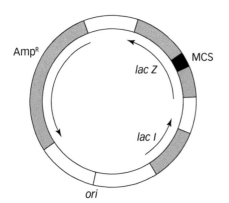

- MCS = Multiple Cloning Site, a region where many restriction sites are grouped

- *lac Z* encodes β-galactosidase, essential for the blue/white screening regime which detects plasmids with a cloned gene

- *lac I* encodes a reversible repressor of *Lac Z*

- Amp^R = ampicillin resistance gene

- *ori* = origin of replication

Figure 14.1 Plasmid cloning vector pUC19 (2686 bp): an efficient, commonly used cloning plasmid created at the University of California.

Vector	Size of DNA that can be inserted	Host cells	Vector details
Plasmids	5–10 kb	Bacteria	Double-stranded circular DNA molecules ➔ See section 6.1 (p.103) for details of plasmids.
Lambda	Up to 23 kb	Bacteria	Bacteriophage
Cosmid	Up to 50 kb	Bacteria	A hybrid lambda/plasmid vector
Bacterial artificial chromosome (BAC)	300 kb	Bacteria	Based on the F1 factor plasmid
Yeast artificial chromosome (YAC)	Up to 3 Mb	Yeast and mammalian cells	Hybrid chromosome with a centromere, telomeres, replication origins, and other key features
Ti plasmid	180 kb	Crop plants	Plasmid of *Agrobacterium tumifaciens*
Retrovirus	8 kb	Human	Modified avirulent virus; used in gene therapy

Table 14.2 Commonly used cloning vectors

Stages in cloning a gene

This section describes the stages in using a plasmid to clone DNA fragments into bacteria.

(i) Nicking DNA

- The plasmid and the DNA to be cloned are cut with the same restriction enzyme: one that that produces staggered ends.
- The overhang enables hybridization of fragment and vector DNA (Figure 14.2).

(ii) Ligation

- The enzyme-nicked plasmid and DNA are mixed in the presence of DNA ligase.
- Complementary base pairing occurs between the cut ends of some of the plasmids and foreign DNA.
- DNA ligase seals the nicks in the sugar–phosphate backbone. A **recombinant** plasmid results.

(iii) Transformation

- Plasmids and bacteria are mixed, at which point the bacteria take up the plasmids.
- Bacteria generally need to be made **competent**, i.e. temporarily permeable to the plasmid by electroporation or chemical treatment.
  ➔ *See section 6. 3 (p. 112) for details of bacterial transformation.*

(iv) Screening

- Bacterial cells are plated on agar medium containing an antibiotic and X-gal.
- This medium selects transformed bacteria, as only those bacteria that have taken up a plasmid, with its antibiotic resistant gene, will survive.

Cloning DNA

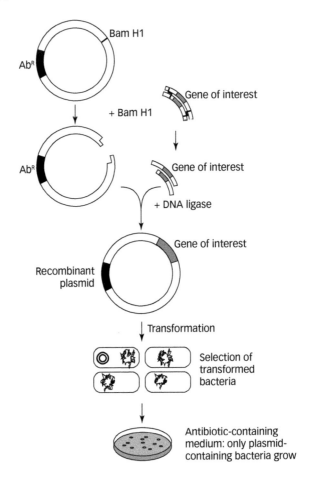

Figure 14.2 Key stages in DNA cloning: producing a recombinant plasmid, transforming bacteria and selecting for plasmid-containing cells. Ab^R=gene for antibiotic resistance.

- Some plasmids will not have the foreign DNA insert (Figure 14.3). X-gal distinguishes bacteria with recombinant or non-recombinant plasmids:
  - white X-gal produces galactose and a blue product when cleaved by β-galactosidase
  - *LacZ* encodes β-galactosidase, and acts as a **reporter gene**, indicating whether or not a bacterial colony contains recombinant plasmids
  - this is achieved because cloned bacteria have been engineered to have half the *LacZ* gene in the bacteria and half in the plasmid
  - in bacteria transformed by *non-recombinant plasmids*, a full functional *LacZ* gene is present—β-galactosidase is produced and X-gal is cleaved; bacteria are blue
  - in bacteria transformed by *recombinant plasmids*, *LacZ* is disrupted—no β-galactosidase is produced, X gal is not cleaved, and so bacteria are white (Figure 14.3).

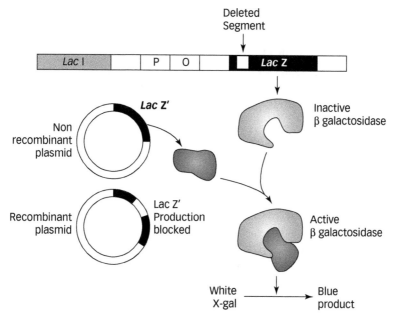

Figure 14.3 Using *LacZ* to identify bacteria with recombinant plasmids.

- Bacterial colonies containing the recombinant plasmid can be removed and grown in liquid culture. This produces millions of copies of the cloned DNA fragment, which can be used in a variety of contexts, e.g. this method provides DNA for genome sequencing projects or to generate transgenic organisms (genetically-modified crops or knockout mice).

Check your understanding

14.1 What roles do (i) restriction enzymes; (ii) vectors; (iii) host cells play in cloning a DNA fragment?

14.2 What are the features that distinguish plasmids, phages, bacterial artificial chromosomes (BAC), and yeast artificial chromosomes (YAC) as cloning vectors?

DNA libraries

Cloning frequently produces a **DNA library**, i.e. a collection of bacterial clones containing DNA from one source.

(i) Genomic DNA libraries

- Purified whole cell DNA is cut into fragments. Each fragment is cloned into a separate bacterium.

Cloning DNA

- The number of resulting clones depends upon:
  - the size of the genome
  - the insert size tolerated by the cloning vector.
- For example, the cloning of the human genome ($\sim 3 \times 10^9$ base pairs) into a cosmid library requires ~350,000 clones to give a 99% chance that every sequence is included in the library (Table 14.1: each cosmid accepts up to 50 kb of DNA).
- **Genomic libraries** are sources of DNA for genomic sequencing projects, for generating transgenic organisms, and for a variety of functional genetics studies.

(ii) cDNA libraries

- cDNA is formed from the mRNA present in a cell.
- mRNA is extracted and converted into complementary DNA (cDNA) using a reverse transcriptase. This cDNA is cloned.
- cDNA is useful in studies of differential gene expression.
- The cDNA produced from genes expressed in a given tissue is termed the **transcriptome**, which contains a mixture of general 'housekeeping' genes and tissue-specific genes.
- cDNA is useful if a project's goal is to express eukaryotic genes in bacterial cells, as cDNA lacks introns and regulatory sequences.
- **Expressed sequence** tags (ESTs) are generated from cDNA libraries. The 5' or 3' end of a cDNA clone is sequenced. This sequence or 'tag' can be used as a probe to find active genes in a particular tissue or at a particular stage of development.
- cDNA **microarrays** are often used to measure expression levels of genes in various tissues or under different conditions (Box 14.1).

Screening DNA libraries

- Genomic and cDNA libraries are commonly screened with a probe to identify gene or DNA fragment of interest.
- A probe is a small known sequence of DNA or RNA (typically about 20 nucleotides) that is complementary to a sequence of interest to which it pairs or hybridizes.
- The nature of the probe depends upon what is already known about the gene of interest. It may be:
  - from the gene sequence of a related species: successful hybridization does not require 100% complementarity between probe and target sequence
  - synthesized from the protein sequence if the protein of the target gene is known.
- Another way of screening a library is to look for a gene's protein product. Clones are tested (in a chemical or antibody test) for the presence of the protein of interest. Note that the DNA library is cloned in an expression vector.

Box 14.1 Using microarrays

- A DNA microarray is a grid of thousands of DNA fragments attached to a solid surface (a glass slide or silicon chip).
- Each DNA spot contains picomolar levels of DNA, which is generally called the probe, and is labelled fluorescently.
- mRNA has been extracted from designated cells and amplified by reverse transcriptase **polymerase chain reaction** to produce cDNA microarrays.
- When looking at differential gene expression:
  - mRNA is extracted from cells under investigation (e.g. at different stages of development or cancerous/non-cancerous cells)
  - extracted mRNA is hybridized to the DNA probes of the microarrays under high stringency conditions
  - the arrays are analysed with an ultraviolet-sensitive camera for fluorescing probes.

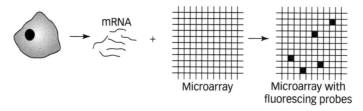

mRNA

Microarray   Microarray with fluorescing probes

Check your understanding

14.3 What is the difference between a genomic and a cDNA library?

14.2 POLYMERASE CHAIN REACTION

- The polymerase chain reaction (PCR) is a second way of producing multiple copies of a specific DNA sequence.
- Starting with a single, or a few, copies of the target DNA, a billion copies can be produced in a few hours.
- The basis of PCR is replication of DNA, catalysed by a DNA polymerase.
- DNA replication requires a single-stranded DNA template and a primer with a 3'OH group to which a DNA polymerase adds new nucleotides.
  - ➡ *See section 7.8 (p. 140) for details of DNA replication.*
- In PCR, two DNA primers, typically 17–25 nucleotides in length, bind to, and define, the target DNA region to be amplified. They also provide the primer for DNA synthesis.
- Thermal cycling is pivotal to PCR. This involves repeated cycles of:
  - heating to denature DNA, and thus provide single-stranded DNA templates
  - cooling for primer annealing and DNA replication.

Polymerase chain reaction

Details of a PCR reaction

- A typical PCR reaction (20–100 µl) includes the following reagents: *Taq* DNA polymerase, DNA primers, target DNA, all four deoxynucleoside triphosphates (dNTPS), and magnesium chloride, in a reaction buffer.
- Figure 14.4 shows the key events in a PCR cycle.
- Each step in a PCR cycle typically lasts 30–60 seconds (the experimenter sets the time).

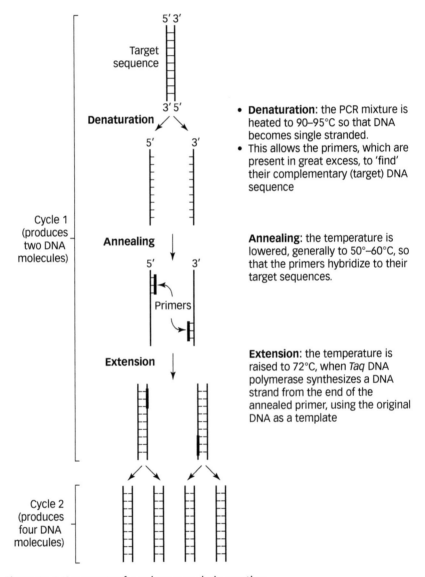

- **Denaturation:** the PCR mixture is heated to 90–95°C so that DNA becomes single stranded.
- This allows the primers, which are present in great excess, to 'find' their complementary (target) DNA sequence

Annealing: the temperature is lowered, generally to 50°–60°C, so that the primers hybridize to their target sequences.

Extension: the temperature is raised to 72°C, when *Taq* DNA polymerase synthesizes a DNA strand from the end of the annealed primer, using the original DNA as a template

Figure 14.4 Key events of a polymerase chain reaction.

- A PCR cycle is repeated 25–35 times; the number of cycles depends on the initial amount of target DNA.
- The amount of DNA doubles each cycle; thus, 1 DNA molecule increases to more than a billion in 30 cycles (2^{30}).
- PCR cycling is automated: it occurs in a **thermocycler** in which metal blocks are programmed to rapidly change temperature.
- Detection and identification of PCR products is by **gel electrophoresis**.
  ➔ *See section 14.3 for details of electrophoresis.*

Looking for extra marks?

Taq DNA polymerase is critical to a PCR reaction. It is derived from the thermobacterium *Thermus aquaticus*, and is thermostable, with a half-life of 40 minutes at 95°C. It has a 5′ to 3′ exonuclease, but no 3′ to 5′ proofreading. Thus, other DNA polymerases, such as *pfu* from *Pyrococcus furiosus*, which possess 3′ to 5′ exonucleases, are used when high fidelity synthesis is needed.

PCR primers

- Well-designed primers are pivotal to successful PCR. Various criteria should be followed.
  - A pair of primers needs to anneal at the same temperature. Thus, when designing primers they must have similar melting temperatures, T_m, the temperature at which 50% of the DNA is denatured. The T is calculated from $4° \times$ (number of G+C in primer) $+ 2° \times$ (number of A+T in primer).
  - The annealing temperature is ~5° below the T_m.
  - Primers should not anneal with other primers in the mixture; otherwise, 'primer dimers' (amplified copies of linked primers) are produced.
  - The primer must not self-anneal, e.g. form loops or hairpins, that disrupt template DNA binding.
  - A maximum GC content of 40–60%.
  - Up to 3 Gs or Cs out of the last 5 bases at the 3′ end of primers increases specificity of primer binding.
- Commercial primer design software packages are available, e.g. 'Primer3' (http://biotools.umassmed.edu/bioapps/primer3_www.cgi).
- **Muliplex PCR** uses multiple primer sets within a single PCR reaction to target multiple genes/target DNA sequences simultaneously. It is cost- and time-efficient, e.g. it is used in forensic DNA fingerprinting.
- In multiplex PCR:
  - annealing temperatures for each of the primer sets must be the same
  - the multiple primer pairs must produce amplicons of varying sizes that form distinct bands when visualized by gel electrophoresis.

- A single primer functions as both a forward and reverse primer in inter-simple sequence repeat-PCR (**ISSR-PCR**) and random amplification of polymorphic DNA (**RAPD-PCR**).
  - ISSR-PCR amplifies DNA between two identical microsatellites. The single primer is complementary in sequence to these flanking microsatellites, e.g. $(GA)_6TG$.
  - RAPD-PCR uses a random ten-nucleotide sequence as both the forward and reverse primer.
- These single primer-based PCRs are often used when no DNA sequence information is available for a species.
- ISSR-PCR and RAPD-PCR produce multiband (typically 3–10) profiles.
- Analysis of the different profiles produced by different individuals has been particularly informative in phylogenetic studies.
- Occasionally, primer specificity is relaxed.
- **Degenerate primers**—mixtures of similar, but not identical, primers—are useful in various contexts, e.g.:
  - if the target sequence is known in a related species
  - primer design is based on protein sequence, as multiple codons can code for a single amino acid.

Limitations of PCR

- PCR requires prior knowledge of the DNA sequence to be amplified so that primers can be designed.
- Rare exceptions include ISSR-PCR and RAPD-PCR primers, making these types of PCR ideal for analysis of previously unstudied species.
- PCR has a high error rate: approximately 1 in 10,000 nucleotides are misincorporated using *Taq* DNA polymerase, which lacks a 3' to 5' exonuclease.
- 'High fidelity' DNA polymerases, such as *Pfu* DNA polymerase, have been engineered with a proofreading exonuclease for more accurate PCR.
- The size of DNA fragment that can be easily amplified by PCR is 2 kb or less. Different polymerases and reaction conditions can extend this to 20 kb or more.
- PCR reactions are easily contaminated. This relates to PCR's efficiency in amplifying extremely small amounts of DNA!
- Primers may bind to non-target sequences. Non-specific amplification can be reduced by:
  - *hot start PCR* when the reaction mix is not active until it is heated to the initial denaturation temperature ($\sim 95°C$), e.g. by using a specialized *Taq* polymerase bound to an inhibitor that only dissociates at high temperature
  - *touchdown PCR* when the annealing temperature of the initial PCR cycles is 5–10°C above optimum; over the first 10–15 cycles, the annealing temperature is lowered 1 or 2°C every 1 or 2 cycles—the higher temperatures give greater specificity for primer binding so that later cycles give more efficient amplification from the target amplicons performed during the initial cycles

○ *nested PCR*—two sets of primers are used in two successive PCRs; if a first PCR reaction generates non-specific DNA products, an aliquot of this DNA can be used in a second PCR reaction using a new pair of primers whose binding sites are 3′ (internal) of each of the primers used in the first reaction.

Real-time PCR

- **Real-time PCR** (RT-PCR) is also known as quantitative RT-PCR; it amplifies and simultaneously quantifies the DNA produced from the target sequence.
- A PCR reaction is prepared as described previously.
- The additional feature of RT-PCR is that the amplified DNA is detected as it forms, i.e. 'in real time'.
- There are two methods to detect the DNA as it forms:
  (i) Non-specific fluorescent dyes that intercalate double-stranded DNA
  (ii) Sequence-specific fluorescent probes.
- A different RT-PCR thermocycler with a special camera is required to monitor fluorescence in each cycle.

(i) RT-PCR with intercalating dyes
- Fluorescent dyes that bind to double-stranded DNA are used, such as SYBER green.
- An increase in PCR product, i.e. as it is formed, leads to an increase in fluorescence intensity. This occurs once per cycle.
- The problem with this quantification method is that the intercalating dyes bind to all double-stranded DNA in the PCR reaction, including non-target DNA, such as primer dimers.

(ii) RT-PCR with a reporter probe
- This type of RT-PCR is also called Taqman PCR.
- The probe anneals to a sequence within the target DNA. It is linked to:
  ○ a reporter dye at the 5′ end
  ○ a quencher dye at the 3′ end, which disrupts the signal from the reporter dye when close to it.
- Figure 14.5 illustrates the use of a Taqman probe.
- This method is more accurate and reliable, but more expensive.

> ### Check your understanding
>
> 14.4 In a typical PCR reaction what processes are occurring at (i) 50–70°C; (ii) 90–95°C; and (iii) 70–75°C?
> 14.5 Can PCR only be used to amplify DNA that has been cloned and sequenced?
> 14.6 What are the advantages of RT-PCR over traditional PCR?

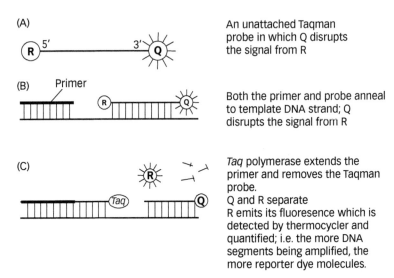

(A)

An unattached Taqman probe in which Q disrupts the signal from R

(B) Primer

Both the primer and probe anneal to template DNA strand; Q disrupts the signal from R

(C)

Taq polymerase extends the primer and removes the Taqman probe.
Q and R separate
R emits its fluoresence which is detected by thermocycler and quantified; i.e. the more DNA segments being amplified, the more reporter dye molecules.

Figure 14.5 Monitoring RT-PCR with a Taqman probe. Q: quencher dye; R: reporter dye.

14.3 GEL ELECTROPHORESIS

- Gel electrophoresis separates DNA and RNA fragments (e.g. PCR products or restriction digests) according to size.
- DNA and RNA segments move under the influence of an electric field.
- Nucleic acids are negatively charged (owing to the sugar–phosphate backbone), so they migrate to the positive anode.
- DNA and RNA fragments move though a gel matrix (commonly agarose or polyacrylamide) at different speeds, according to their size and structure.
- Agarose efficiently separates larger DNA molecules, e.g. PCR products.
- Polyacrylamide gives high resolution of short DNA molecules of a similar size, e.g. it is used in DNA sequencing.
- Agarose gels are generally run in a horizontal 'slab' format (Figure 14.6).
- Polyacrylamide gels are more often used in a vertical capillary format.
- Dyes, such as ethidium bromide (visualized under ultraviolet light), bind DNA so it can be located in a gel.
- Separated DNA molecules can be sized by running **DNA ladders**, which are mixtures of DNA fragments of known size.
- Fragment size measurement and analysis are often performed with a specialized gel analysis software.
- Capillary electrophoresis results are typically displayed as a 'trace' or electropherogram (see Figure 14.8).
- DNA can be sliced out from a gel and purified for future work.

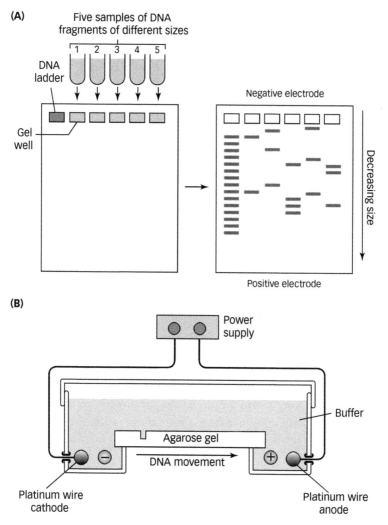

(A)

Five samples of DNA
fragments of different sizes

1 2 3 4 5

DNA ladder

Negative electrode

Gel well

Decreasing size

Positive electrode

(B)

Power supply

Buffer

Agarose gel

DNA movement

Platinum wire cathode

Platinum wire anode

Figure 14.6 Principles of agarose gel electrophoresis. (A) Separation of DNA fragments (e.g. after 100 V for 1 h); (B) key features of an electrophoretic tank.

14.4 DNA SEQUENCING

DNA sequencing determines the order of nucleotide bases in a DNA molecule.

Chain termination method

- This is the standard sequencing method developed by Frederick Sanger and collaborators in 1977.
- Multiple copies are made of the DNA whose sequence is required.
- These copies incorporate different kinds of nucleotides, **dideoxynucleotides**, which are labelled fluorescently and detected by the sequencing machine,

DNA sequencing

- The reaction that produces DNA for sequencing is termed **cycle sequencing**. It is essentially a PCR reaction, *but* copies are made of only one DNA strand.
- A cycle sequencing reaction reaction includes a *Taq* DNA polymerase, a single DNA primer, template DNA, and dNTPs.
- The primer is complementary to the 5′ end of the DNA to be sequenced. The sequence of the 5′ end is generally known as it is the edge of the vector into which the DNA was cloned or the primer sequence of a PCR amplicon.
- Each dNTP is a mixture of the 'normal' deoxynucleotides and a small fraction of fluorescently labelled dideoxyribonucleoside triphosphates (**ddNTPs**).
- ddNTPs lack the 3′ OH on the deoxyribose sugar; when incorporated into a growing DNA chain they terminate DNA synthesis.
- Cycle sequencing involves 25–30 cycles of denaturation, annealing, and extension.
- During the extension step, *Taq* polymerase randomly adds either a dNTP or a ddNTP to the growing chain.
- Whenever a ddNTP is added elongation stops.
- Thus, when cycle sequencing is complete, there are PCR fragments of all possible different lengths, each terminated by a ddNTP (Figure 14.7).
- The PCR fragments are analysed by an automated DNA sequencer.
  - The different sized fragments are separated by electrophoresis in a tiny capillary tube.
  - This tube then passes across an electronic camera, which captures an image of each band and its fluorescence intensity.
  - The results are presented as the characteristic sequencing trace (Figure 14.8).
  - Each ddNTP has a different coloured fluorophore attached, so the four different bases can be identified. A is green, T is red, G is black, and C is blue.

```
ATCCGGATGCAATGTCACTA
ATCCGGATGCAATGTCACT
ATCCGGATGCAATGTCAC
ATCCGGATGCAATGTCA
ATCCGGATGCAATGTC
ATCCGGATGCAATGT
ATCCGGATGCAATG
ATCCGGATGCAAT
ATCCGGATGCAA
ATCCGGATGCA
ATCCGGATGC
ATCCGGATG
ATCCGGAT
ATCCGGA
ATCCGG
ATCCG
ATCC
ATC
AT
```

Figure 14.7 The set of PCR fragments produced during cycle sequencing of a 20 nucleotide DNA fragment. Letters in bold font represent the dideoxynucleotides.

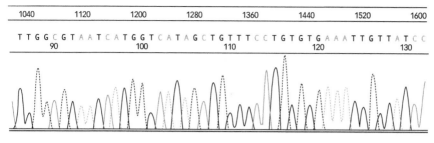

Figure 14.8 A sequencing trace.

- ○ The shortest sequencing fragment corresponds to the length of the PCR primer plus one nucleotide.
- In one sequencing reaction just one DNA strand is sequenced.
- The sequence of the opposite strand can be deduced through complementary base pairing.
- If accuracy of the first sequencing reaction is in doubt then the complementary strand can also be sequenced.
- The traces for the first 15–40 nucleotides are often of poor quality.
- A maximum of 500–900 nucleotides can be sequenced in one reaction.
- Long pieces of DNA are sequenced by amplifying small overlapping fragments, which are sequenced separately.

Whole genome sequencing

- There are two main approaches to sequencing genomes: the **clone-by-clone** and the **shotgun** approach.
- Both approaches involve:
  - ○ chopping genomic DNA into fragments with restriction enzymes
  - ○ cloning these fragments
  - ○ sequencing these cloned fragments separately
  - ○ assembling the sequenced fragments into the complete genome sequence.
- The two methods differ in the size of cloned fragment.
  - ○ *Shotgun approach*: ~1-kb fragments are cloned into plasmids. This sized fragment is sequenced directly.
  - ○ *Clone-by-clone approach*: 100–150 kb fragments are cloned into BACs or YACs. These large inserts have to be further fragmented to sequence.
- The critical factor in ordering the sequenced DNA fragments is that their sequence **overlaps** (Figure 14.9).
- The continuous genomic sequence is deduced by assembling the overlapping sequenced DNA fragments into **contig maps**.
- A contig (contiguous) map is a set of overlapping DNA fragments that have been assembled in the correct order to form a continuous stretch of DNA.

DNA sequencing

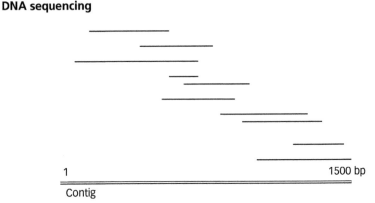

Figure 14.9 Producing a contig map: a set of ten DNA fragments have been analysed by a computer 'assembly' programme, which pieced them together into a contig.

- A contig is an abbreviation for 'contiguous sequenced region'.
- Positional genetic mapping then places the DNA sequence in its chromosomal position.

  ➔ *See section 4.6 (p. 77) for details of positional gene mapping.*
- There are advantages and disadvantages of each approach to genomic sequencing.
  - ○ *Shotgun sequencing* is fast, but it is difficult to assemble so many tiny DNA fragments into their correct order. Powerful computers produce the contig maps.
  - ○ *Clone-by-clone sequencing* is more reliable but slow, especially the mapping step.
- The human genome was sequenced by a combination of both methods.
- New methods of sequencing are being developed, as reviewed in the next section.

Next generation sequencing

- New techniques are being developed to make sequencing faster and cheaper (**next generation sequencing**).
- These aim to sequence hundreds of thousands of DNA fragments simultaneously.
- During **pyrosequencing**, a nucleotide sequence is determined as it is synthesized.

Principles of pyrosequencing
- Genomic DNA is cut by restriction enzymes into 300–800 bp fragments.
- **Adapters**—short oligonucelotides—are ligated to each end of the fragments. Adapters provide the priming sequence for PCR amplification.
- DNA is made single stranded and each fragment is attached to a primer-coated bead and captured within a water droplet.
- Each DNA droplet is forced into a single well within a sequencing microarray and mixed with sequencing reagents, *but* excluding dNTPs.
- Nucleotides are added to the plates in turn. If the template DNA (i.e. DNA fragment being sequenced) specifies addition of that particular nucleotide the sequencer detects this addition.

- The **pyrophosphate (PP$_i$)**, cleaved from the dNTP as it is added, is used in a chemical luminescent reaction (the reagents for this are in the sequencing mix in the microarray well). The generated light is detected by the sequencer (Figure 14.10).
- Unincorporated nucleotides are degraded by apyrase and another nucleotide batch is added.
- The four nucleotides are added in repeated succession into the wells and the nucleotides incorporated recorded.

Analysing DNA sequences

- DNA sequencing is performed by researchers in many different fields of biology.
- The resulting sequences can be analysed by a wide range of computer programmes.
- **Bioinformatics** refers to the computational analysis of nucleic acid and protein sequence.

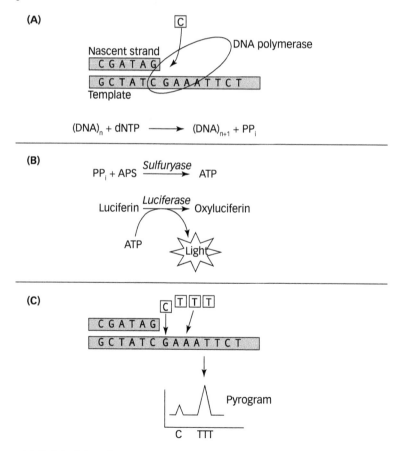

Figure 14.10 Principles of pyrosequencing.

DNA sequencing

- Many bioinformatic programmes can be accessed and downloaded from the World Wide Web.
- Internet sequence databases, e.g. GenBank, are an invaluable resource for comparative information.

(i) Sequence alignment

- Relationships between sequences are assessed by aligning them, either in a **pairwise** or **multiple** alignment.
- Similarity scores are assigned to the alignments.
- BLAST is a much-used pairwise alignment tool.
- Multiple alignments are commonly performed with Clustal Omega or MAFFT.
- BLAST is commonly used to identify a sequence of interest.
  - It compares a new sequence to all known sequences in a database, such as GenBank.
  - Matches to sequences in the database are ordered, starting with the closest matches.
- BOLD or DNA barcoding for life is an integrated species identification database.
- It uses a short DNA sequence of the:
  - mitochondrial cytochrome oxidase I (*COI*) gene in animals
  - chloroplast maturase K (*matK*) or ribulose biphosphate carboxylase (*rbcL*) in plants
  - nuclear **internally transcribed spacer** (ITS) region in fungi.
- Aligned sequences can be used in comparative genomics to explore evolutionary relationships between taxa.
- PAUP and PHYLIP are just two of a wide range of phylogenetics computer programs that can be used to produce phylogenetic trees.

(ii) Gene prediction

- The identification of regions of DNA sequence that encode genes is an important part of sequencing the genome of a species.
- Programmes, such as GenScan, analyse DNA sequence for **open reading frames** (ORFs), i.e. a continuous DNA sequence that contains a start and stop codon in the same reading frame.
- ORFs are assumed to code for a polypeptide.
- Other sequence indicators of a protein encoding gene include splice recognition sites (in eukaryotes) and nearby potential promoter elements.
- There are various programmes that can predict resulting protein three-dimensional structure from an ORF.

Check your understanding

14.7 How are ddNTPs used in the chain termination method of DNA sequencing?

14.8 How does shotgun sequencing differ from the clone-by-clone approach?

14.5 FORWARD AND REVERSE GENETICS: FUNCTIONAL GENOMICS

- The ultimate goal of genetics is to understand the function of genes.
- There are two contrasting approaches to investigating gene function, i.e. to linking genotype and phenotype.

  (i) **Forward genetics** (also termed the 'top-down' approach). This starts with a phenotype (mutant organism) and the gene encoding it is deduced.

  function (phenotype) - - - - - - - - - - - -> gene

  (ii) **Reverse genetics** (also termed the 'bottom-up' approach). This starts with a genotype (a DNA sequence) and its function (phenotype) is deduced.

  gene - - - - - - - - - - -> function (phenotype)

Forward genetics

- This is the historical/traditional approach. It involves:
  - introducing mutations into a population
  - screening for the existence or lack of a phenotype
  - mapping relevant gene loci through genetic crosses or other means, such as positional cloning (depends on the species involved)
  - isolating the relevant gene(s) and DNA sequencing.
- Because mutation rates are low, large numbers of organisms are needed to screen for interesting phenotypes. Thus, the ideal species:
  - has a short generation time
  - produces a large number of progeny
  - is easy to rear
  - is easy to score phenotypically
  - has high resolution genetic and physical maps, and can be manipulated genetically.
- In the early twentieth century forward genetics was particularly effective with bacteria, yeast (*Saccharomyces cerevisiae*), and *Drosophila melanogaster*. More recently, geneticists have adopted this approach with the nematode *Caenorhabditis elegans* and zebrafish.
- The choice of mutagen (radiation, chemical, or transposable elements) is important as different mutagens cause different types of mutations.
- For example, certain chemicals cause base pair substitutions, while others cause insertions/deletions with differing effects on phenotype. The latter are more likely to completely abolish function.

  ➤ *See section 11.3 (p. 215) for details of mutagens.*
- Mutagen dose is critical: you need enough mutants for analysis, but too high a mutagen dose might cause multiple mutations in an individual making it difficult to identify genes responsible for a mutant phenotype.

Forward and reverse genetics: functional genomics

➔ *See section 6. 3 (p. 106) for techniques of producing and screening bacterial mutants.*

- Mutations in multiple genes often produce the same phenotype, so **complementation tests** are performed to check whether interesting mutants are in the same or different genes, i.e.:
  - two mutants (homozygotes) are crossed and the F1 is analysed
  - if the F1 expresses the wild type phenotype, we conclude each mutation is in one of two possible genes necessary for the wild type phenotype.
- One highly informative mutant screen of recent decades was Christine Nusslein-Volhard, Eric Weishaus, and Ed Lewis's production and analysis of *Drosophila* developmental mutants. They:
  - treated thousands of flies with mutagens
  - spent a year microscopically examining embryos for developmental anomalies.
- Nusslein-Volhard and Weishaus identified 15 different early embryo mutants whose analysis gave insights into mechanisms that establish the arthropod segmented body plan.
- They identified various **gap**, **pair-rule**, and **segment polarity** genes.
- Lewis's work identified **homeotic** genes: genetic switches that turn on and off different programmes of cellular differentiation.
- Homeotic genes, remarkably conserved in type and sequence, have been subsequently identified in numerous other vertebrate and invertebrate species.

Reverse genetics

- This is a molecular approach. Starting with the gene of interest, it involves:
  - mutating the gene (**site-specific mutagenesis**)
  - using a vector to introduce the mutated gene *in vitro* into a fertilized egg or embryonic stem cells where the mutated gene integrates into the host cell's genome randomly or in a targeted way
  - introducing transformed cells into an experimental organism and examining resulting phenotypes.
- Reverse genetics produces **transgenic** organisms, i.e. an organism whose genome contains foreign DNA.
- Transgenic mice have been created frequently in the study of human gene function. Many other transgenic animals (e.g. sheep, pigs, flies) and plants (e.g. tobacco, wheat) have been engineered.

Site-specific mutagenesis

- The gene or DNA sequence to be mutated is cloned into a circular vector and mutated as shown in Figure 14.11.

(A)

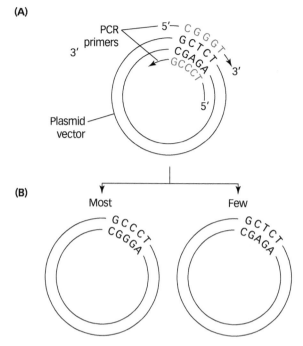

Figure 14.11 Principles of site-specific mutagenesis.

Producing a transgenic organism

- Recombinant DNA technology is first used to build the DNA molecule to insert into an experimental organism. The molecule typically contains:
  - the gene whose function is being investigated
  - vector DNA (e.g. plasmid or retrovirus) to insert the gene into the host cell's genome
  - promoter and enhancer sequences, which enable expression of the gene in the host cell.
- The DNA construct is then used to produce a transgenic organism by either:
  - (i) transforming embryonic stem cell (ES cells)
  - (ii) injecting newly-fertilized eggs.

(i) ES cell method

- ES cells are harvested from the inner cell mass of mouse blastocysts and grown in culture.
- They are transformed with the vector containing the gene of interest.
- Cells that have taken up and integrated the gene into their genome (Box 14.1) are injected into the inner cell mass of the mouse blastocyst.
- This blastocyst is transferred into the uterus of a pseudopregnant mouse (a female that was mated with a vasectomized male).

- Any resulting transgenic offspring will be heterozygous. The presence of the gene of interest can be identified by molecular analysis, e.g. by removal of a small piece of tail or ear tissue, extracting DNA, and carrying out PCR and/or sequencing.
- Mice that are heterozygous for the desired gene are mated together.
- One in four of their offspring will be homozygotes for the gene. Interbreeding these mice produces the transgenic strain.
- Having produced homozygous mice with the mutant phenotype the role of the inserted gene can now be investigated.

(ii) Pronucleus method
- Newly fertilized eggs are harvested and those with their pronuclei still separate are used.
- The gene of interest is injected into one of the pronuclei.
- Eggs with the gene of interest integrated into their genome (Box 14.2) are selected, grown for a few generations, and implanted into a pseudopregnant female.
- Identification of transgenic mice proceeds as for ES cells.

Knockout mice
- Sometimes the mutated version of the inserted gene abolishes function of the encoded protein. The resulting homozygous transgenic mice are termed **knockout mice**.
- About 15% of gene knockouts are developmentally lethal, i.e. although expressed in specific adult tissues, they are vital embryologically.
- These genes can be studied in adult mice using the *cre/loxP* expression system, which produces **tissue-specific knockout mice.**
- Cre is a recombinase produced by *Escherichia coli* phage P1 that recognizes a particular sequence termed *loxP*. DNA between two *loxP* sites is removed and the remaining DNA ligated.
- To produce tissue-specific knockout mice the following are inserted into the mouse genome:
  - the *cre* gene with a tissue-specific promoter
  - the target gene flanked by *loxP* sequences.
- Activation of the *cre/loxP* system occurs in the relevant tissue in the following way:
  - cells receive a signal, e.g. hormone or cytokine
  - this stimulates production of tissue specific transcription factors to turn on tissue-specific genes and also the *cre* gene
  - the resulting *cre* protein binds to the *loxP* sequences and removes the target gene.
- Thus, the phenotype of a mouse with the specific gene knocked out in only certain cells can be studied.

Box 14.2 Targeted gene insertion

- The vector contains:
  - the gene of interest plus known flanking sequences
  - *neor*, which encodes an enzyme to inactivate neomycin-type antibiotics, e.g. drug G418—this identifies cells with the vector
  - *tk*, which encodes thymidine kinase—this phosphorylates the guanine nucleoside analogue ganciclovir, which, when incorporated during DNA replication, terminates DNA synthesis; it is used to identify cells that have randomly incorporated the gene of interest.
- During production of a transgenic organism:
  - the vector is added to a culture of ES cells or injected into an egg
  - cells are treated with G418 and ganciclovir
  - G418 kills cells (the majority) which failed to take up a vector
  - ganciclovir kills cells (a few) in which the vector randomly inserted in the genome.

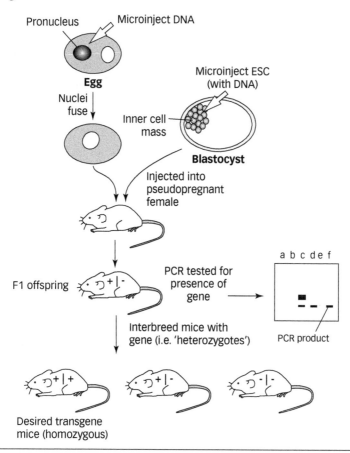

Forward and reverse genetics: functional genomics

- The *cre/loxP* system can also be used to produce **knock-in mice**. *Cre* can be activated to remove sequences that block gene transcription, e.g. a stop sequence between the promoter and gene. Thus, a gene can be turned on in certain cells or at certain times.

Knockdown mice

- The creation of transgenic organisms permanently alters their phenotype.
- The recent discovery of **RNA interference** gives a means of temporarily switching off expression of a gene and analysing the phenotypic consequences.

  ➔ *See section 10.4 (p. 204) for details of types of RNA that cause RNA interference.*
- Double-stranded RNA is synthesized with a sequence complementary to a gene of interest and is introduced into a cell or organism.
- It is recognized as exogenous genetic material and activates the RNAi pathway:
  ○ dicer enzymes cleave the dsRNA into siRNAs
  ○ these are picked up by the RNA-induced protein complexes (RISCs).
  ○ RISCs coordinate binding of siRNAs with target mRNA, thus silencing expression of the gene of interest.
- RNAi may not totally abolish expression of the gene, thus this approach is sometimes termed **knockdown** to distinguish it from 'knockout' transgenic organisms in which expression of a gene is eliminated entirely.
- The greatest challenge with using RNAi to probe gene function is introducing dsRNA into cells.
- In cell culture dsRNA can be injected directly into cells.
- DNA vectors can be used to introduce the dsRNA into cells or organisms. For example *E. coli*, containing a dsRNA expression vector, are fed to *C. elegans*.

Check your understanding

14.9 What are the main stages in producing a transgenic mouse?

14.10 What are the main differences between the forward and reverse genetics approaches to identifying the function of particular genes?

online resource centre You'll find guidance on answering the questions posed in this chapter—plus additional multiple-choice questions—in the Online Resource Centre accompanying this revision guide. Go to **www.oxfordtextbooks.co.uk/orc/thrive/** or scan this image:

Glossary

A-DNA A form of DNA that occurs in dehydrated samples. It is a compacted right-hand helix.

Acceptor arm The site of amino acid attachment on a tRNA molecule. It contains the base-paired 5′ and 3′ ends of the tRNA. The amino acid binds to the CCA overhang on the 3′ strand.

Acentric A chromosome without a centromere.

Acrocentric A chromosome with the centromere located close to one end.

Activator A protein that regulates the initiation of transcription through interaction with RNA polymerase and upstream enhancer sequences.

Addition rule The probability of one event or another event occurring is calculated by adding the probabilities of each separate event.

Additive genes Genes whose alleles make an approximately equal, quantifiable contribution to phenotype.

Additive genetic variance The component of genetic variation that makes the major contribution to resemblance between related individuals and thus the response to selection by a group of individuals.

Adenine A purine nitrogenous base which pairs with thymine in DNA and uracil in RNA.

Affected sibling pair analysis A technique for identifying quantitative trait loci (QTLs). A large number of sibling pairs that both express a quantitative condition are scanned for shared alleles of polymorphic markers: shared alleles indicate marker linkage to a relevant QTL.

Alkylating agent A chemical that replaces hydrogens within DNA with an alkyl group.

Alleles Alternative forms of a gene or other DNA sequence.

Allele frequency The proportion of a given population carrying that allele.

Allopolyploidy Polyploidy that results from cross species hybridization: chromosomes are derived from two or more species.

Alphoid DNA The tandemly repeated satellite DNA of a centromere.

Alternation of generations The division of a life cycle into a distinct haploid and diploid phase.

Alternative splicing The process by which exons of pre-mRNA are joined in different ways during RNA splicing to produce different mRNAs translated into different protein isoforms.

Ames test A biological assay to assess the mutagenic potential of chemical compounds. It assesses a compound's potential to mutate bacteria.

Amino acid Monomer building blocks linked by peptide bonds to form proteins.

Amino acyl (A) site The site of a ribosome that accepts the incoming aminoacyl tRNA during translation.

Amino acyl synthetase An enzyme that catalyses charging of a tRNA with its amino acid.

Anaphase A stage in mitosis and meiosis when chromatids or chromosomes move apart to opposite poles of the cell.

Aneuploidy Chromosomal condition in which single chromosomes are missing or an additional copy present.

Glossary

Anticipation The expression of an age-dependent condition at earlier ages and with greater severity in succeeding generations.

Anticodon The three-base codon of tRNA that is complementary to an mRNA codon.

Antiparallel The situation in which the two polynucleotide strands of the DNA double helix run parallel to each other, but with opposite alignments.

Antisense strand This is the template DNA strand during transcription. It has the complementary sequence to the mRNA.

Apurinic site This is a site in DNA that has lost its purine base spontaneously or as a result of DNA damage. If unrepaired a mutation may occur.

Apyrimidinic site This is a site in DNA that has lost its pyrimidine base.

Archaea A group of single-celled microorganisms, distinct from bacteria, that lack membrane-bound organelles and often live in extreme environments.

Association study A population-based study that examines whether a particular allele or DNA marker co-occurs with a phenotype (e.g. a disease) at a significantly higher rate than predicted by chance alone.

Assortative mating Mating in which partners are more phenotypically or genetically similar to each other than would be expected with random mating.

Aster A star-shaped arrangement of microtubules radiating from each centriole during mitosis.

Attenuation A transcription regulation mechanism. Slowed transcription through a regulatory region causes it to form a hairpin, which terminates transcription.

Autopolyploidy Polyploidy that results from duplication of all a species' chromosomes.

Autosome Any chromosome other than a sex chromosome.

Auxotroph A mutant bacterium of fungus that requires a particular growth nutrient not required by the wild type strain.

B-DNA The main form of DNA. It is a right-handed helix of diameter 23.7 Å.

Backcross The cross of an F1 individual with one of its parents.

Bacterial artificial chromosome (BAC) A cloning vector, based on the F plasmid, into which large segments of DNA (100–300 kb) can be inserted.

Bacteriophage A virus that infects bacteria.

Balanced selection Selection that maintains phenotypes of intermediate expression.

Balanced translocation A translocation in which no chromosomal material has been lost or gained, or gene function disrupted.

Barr body This is a condensed inactive X chromosome in the nucleus of a somatic cell.

Base excision repair (BER) A DNA repair system initiated by a DNA glycosylase removing the damaged base.

Binary fission A form of asexual reproduction in which a cell divides into two identical ones.

Bioinformatics The use of computers to store, visualize, and analyse biological data.

Bivalent Paired homologous chromosomes during prophase I and metaphase I of meiosis.

Blunt ends A DNA fragment that has no single strand overhangs at either of its two 5′ or 3′ ends.

Bottleneck A period of low population numbers when genetic diversity is lost.

Broad sense heritability (H²) The proportion of phenotypic variance that is due to total genetic variance.

cDNA This is DNA synthesized from an mRNA template using reverse transcriptase and DNA polymerase.

C value The size of a genome stated as number of base pairs.

C value paradox The lack of a consistent relationship between the C value and the complexity of an organism.

Candidate gene A gene suspected of being the gene responsible for a disease or other phenotype under investigation.

Cell cycle The sequence of events between one cell division and the next.

Cell cycle checkpoints Control points in the cell cycle that check that a cell is ready for S and M phases.

Cell plate A membrane-bound space that forms in the middle of a dividing plant cell and later becomes the cell wall.

CentiMorgan (cM) A unit for measuring the distance between linked genes, derived from recombination frequencies.

Central dogma The idea that DNA makes RNA which makes protein.

Centrioles A minute pair of organelles, close to the nucleus, that make the spindle fibres during mitosis and meiosis.

Centromere The part of a chromosome that attaches to the spindle fibres during mitosis or meiosis.

Chaperone protein A protein that aids correct tertiary folding of a newly synthesized polypeptide.

Chiasma(ta) The region of contact between non-sister chromatids during pairing of homologous chromosomes at prophase I of meiosis. Exchange of homologous chromatid sections occurs at a chiasma.

Chromatid One of two identical DNA molecules formed by duplication of a chromosome.

Chromosome An organized package of DNA and proteins found in the nucleus of a cell, which contains the genes. Chromosomes come in pairs, with one derived from each parent. Each species has a specific number of chromosomes.

Co-dominance The situation when both genes of a pair in a heterozygote are fully expressed, with neither one being dominant or recessive to the other.

Codon A sequence of three nucleotides along a DNA or RNA molecule, which represents an amino acid.

Coefficient of coincidence A measure of crossover interference caused by a nearby crossover.

Colchicine An alkaloid extracted from the autumn crocus which inhibits spindle microtubule assembly, and so arrest cells in metaphase of mitosis.

Col plasmid A plasmid that produces toxins that kills non-host bacteria.

Competent cell A cell able to take up DNA from the environment.

Complementary base pairing The pairing of adenine with thymine and guanine with cytosine.

Complementary genes A full phenotype is only produced when at least one dominant allele of each gene is present.

Complementation test A test to determine whether two recessive mutations are in the same or different genes.

Compound heterozygote A homozygous recessive individual with two different recessive alleles.

Concordance The presence of the same trait in a pair of twins.

Conjugation Transfer of a plasmid from one bacterium to another via a pilus, a bridge-like protein extension.

Consanguineous mating Mating between biological relatives.

Glossary

Consensus sequence A DNA sequence that is identical or similar in different species. It often has functional significance.

Constitutive gene A gene, commonly required for basic cellular metabolism or architecture, that is expressed (transcribed) continuously.

Constitutive heterochromatin Chromosomal regions that are in a permanent compacted non-expressed state.

Contig A set of overlapping DNA fragments that is used to deduce the sequence of a DNA segment.

Continuous trait A trait that is measured and which has a large number of possible phenotypes between two extremes.

CpG islands DNA segments with an increased density of the dinucleotide sequence CG; commonly found in promoters.

Crossing over An exchange of genes between two non-sister chromatids of a homologous pair of chromosomes.

Cycle sequencing The polymerase chain reaction that produces DNA for sequencing.

Cyclin Proteins that control a cell's progression through the cell cycle by activating cyclin-dependent kinases.

Cytokinesis The division of the cytoplasm into two daughter cells at the end of mitosis or meiosis.

Cytosine Pyrimidine nitrogenous base that pairs with guanine in DNA and RNA.

Deamination The loss of an amine group which converts cytosine to uracil and adenine to hypoxanthine.

Degeneracy Many amino acids are specified by two or more different codons.

Deletion The loss of genetic material. The size of the loss varies from a single nucleotide to a segment of a chromosome.

Denaturation The separation of double-stranded DNA into single strands by heating.

Deoxyribose A pentose sugar that is a constituent of deoxyribonucleotides.

Depurination The loss of a purine base from a nucleotide.

Derivative chromosome (der) A structurally rearranged chromosome generated by a chromosome mutation involving two or more chromosomes; typically the result of a translocation.

Dicentric A chromosomes with two centromeres.

Dicer An endo-ribonuclease that cleaves precursor double-stranded RNA into siRNA and miRNA.

Dideoxyribonucleoside triphosphate (ddNTP) Nucleotides that lack a 3'-OH group on their deoxyribose sugar. They terminate a growing nucleotide chain.

Dihybrid cross A genetic cross that analyses the inheritance patterns produced by the segregation of alleles of two genes.

Diploid A cell that contains two sets of chromosomes.

Directional selection Selection in which individuals at one end of the normal distribution of a continuous trait are at an advantage over other individuals.

Disassortative mating A mating preference for individuals with a genotype or phenotype different to one's own.

Discontinuous trait A trait that is inherited in a Mendelian fashion and occurs as two or three distinct phenotypes.

Disruptive selection Selection in which individuals at both extremes of the normal distribution of a continuous trait are at an advantage over individuals of intermediate phenotype.

Dizygous twin A twin derived from two separate fertilization events.

DNA cloning The transfer of a DNA fragment from one organism to a self-replicating genetic element that replicates the fragment in a foreign host cell.

DNA fingerprinting The production of DNA profiles that are unique to individuals.

DNA glycosylase An enzyme that recognizes and excises a damaged nitrogenous base from DNA.

DNA gyrase An enzyme that unwinds DNA.

DNA helicase An enzyme that separates the double helix into single strands.

DNA ladder A mixture of DNA fragments of different lengths used in gel electrophoresis as a size reference.

DNA library A cloned collection of DNA fragments from one source.

DNA ligase An enzyme that joins the 5′ end of one polynucleotide chain to the 3′ end of another polynucleotide chain.

DNA marker A DNA sequence that is used experimentally to identify a sequence, gene, chromosome, or individual.

DNA polymerase An enzyme that catalyses the synthesis of DNA in a 5′ to 3′ direction using a template DNA molecule, a primer, and free deoxyribonucleotides.

DNA repair Various processes by which a cell identifies and corrects damage to DNA.

DNA replication A process during which each strand of a DNA molecule serves as a template for synthesis of a new complementary strand.

Dominance The situation where one member of an allele pair is expressed to the exclusion of the other.

Dominant epistasis The situation when expression of alleles at one locus is suppressed in individuals which are have a dominant allele at a second, epistatic locus.

Dosage compensation The random inactivation of one X chromosome in each cell of a female, so that males and females produce equal levels of proteins encoded by X-linked genes.

Double crossover Two separate crossover events occurring between the same pair of homologous chromosomes.

Duplication The presence of one or more extra copies of a gene or chromosome segment.

Elongation factor A protein required for polypeptide elongation during translation.

Endonuclease An enzyme that cleaves a polynucleotide at internal positions.

Endo(poly)ploidy A situation in which certain tissues within an organism have cells with extra sets of chromosomes. It occurs when chromosomes replicate without subsequent nuclear division.

Endosymbiotic theory The theory that mitochondria and chloroplasts originated through prokaryotes being taken inside another prokaryote by endosymbiosis.

Enhancer A DNA control sequence, often located tens of kilobases upstream from a gene, that has binding sites for activators and repressors.

Environmental variance The proportion of phenotypic variance that is due to differences in the environmental factors acting on individuals in a population.

Epigenetics Inherited changes in phenotypic expression that are caused by mechanisms other than changes in DNA sequence.

Episome A circular piece of DNA that can replicate independently of the bacterial chromosome or integrate and replicate as part of the chromosome.

Glossary

Epistasis Interaction between genes, such that certain alleles of one gene mask or interfere with expression of the other gene.

Equator The midpoint between two centrioles in a cell undergoing mitosis or meiosis where pairs of chromatids or homologous chromosomes align during metaphase.

Eubacteria Unicellular prokaryotic organisms.

Euchromatin Chromosomal regions that are relatively uncoiled during interphase and contain actively transcribed genes.

Eukaryote Any organism with distinct membrane-bound organelles.

Exit (E) site The site of a ribosome containing a tRNA after it has donated its amino acid to the growing peptide chain.

Exon A DNA segment that contains information.

Exon shuffling Evolution of new genes by combining exon from two or more pre-existing genes.

Exonuclease An enzyme that removes successive nucleotides from the end of a polynucleotide chain.

Exosome Multi-protein RNase-containing complexes found in the nucleus and cytoplasm whch degrade incompletely processed RNA molecules.

Expressed sequence tag A short cDNA fragment of an expressed sequence which serves as a landmark for gene mapping.

Expression vector A cloning vector which contains the necessary control sequences for expression of its cloned gene.

Expressivity Variations in phenotype among individuals with the same genotype.

F1 (first filial) generation The first generation produced from mating parents showing distinct phenotypes and genotypes.

F2 (second filial) generation The generation produced by mating two individuals from the F1 generation.

F plasmid A plasmid that controls conjugation between bacterial cells.

Facultative heterochromatin Compaction of this chromatin varies according to cell type and stage of development.

Fitness The ability of an organism to transmit its genes to the next generation.

Fixation Situation in which an allele has spread to all members of the population that carries it. NB, the only allele.

5′ Cap A special nucleotide (7′-methylguanine) at the 5′ end of mRNA that helps to maintain stability of the molecule.

5′ Splice site The sequence found in a pre-mRNA which is cut to free the 5′ end of the intron during splicing.

5′ Untranslated region (UTR) A sequence of ribonucleotides at the 5′ end of mRNA that does not contain a protein coding sequence, but influences the translation efficiency and the stability of the mRNA.

Fluorescent in situ hybridization A technique to detect and localize the presence or absence of specific DNA sequences on chromosomes. It uses fluorescent probes that bind to relevant sequences and are detected by ultraviolet microscopy.

Formylmethionine A form of methionine that is the first amino acid incorporated into a polypeptide chain during prokaryotic translation.

Forward genetics One approach to identifying a gene's function: starting with a particular phenotype the encoding gene is deduced.

Founder effect The drift-induced change in the genetic constitution of a population that was initiated from a small number of individuals in a novel habitat (the founders).

Fragile site A heritable position, often containing an expanded tri-nucleotide repeat, on a chromosome that is prone to breakage when a cell is exposed to various stresses.

Frameshift mutation The insertion or deletion of one or two nucleotides into a coding DNA sequence which changes the reading frame.

G_0 The state of fully differentiated cells that have stopped cycling.

G_1 The first, and main, stage of a cell cycle when cells are growing and carrying out their normal functions.

G_2 The stage of the cell cycle following DNA replication.

Gametophyte In plants, the haploid gamete-producing generation.

Gel electrophoresis A technique that separates segments of DNA and RNA and polypeptides on the basis of size and differences in electrical charge of the molecules.

Gene The fundamental unit of heredity that consists of a segment of DNA at a specific locus on a chromosome which determines a particular trait of an organism.

Gene family A set of related genes, with similar, but not identical, sequences that code for similar biochemical functions.

Gene flow Genetic exchange between populations resulting from the migration of individuals followed by sexual reproduction.

Gene pool All genes and their alleles carried by the reproductive members of a population.

Genetic code All 64 codons and the amino acids they encode (including the three stop codons).

Genetic drift Random changes of the frequency of alleles in succeeding generations of a population. It results from unequal sampling of gamete types during fertilization.

Genetic map A linear representation of a chromosome showing the positions and relative distances between genes and other chromosomal landmarks.

Genetic variance The proportion of phenotypic variance that is due to differences in genotype between individuals in a population.

Genome A species' haploid DNA content, i.e. the genetic content of a single set of chromosomes.

Genomic imprinting A pattern of gene expression that is determined by the parental origin of an allele.

Genomic library A collection of bacterial clones that represent a species' whole genome.

Genotype The genetic make-up of an individual with respect to one genetic locus, a number of loci, or its total genetic content.

Genotype frequency The proportion of a given population carrying a given genotype.

G/M checkpoint Control mechanism that prevents cells starting mitosis with unreplicated DNA.

Goodness-of-fit chi-squared test A statistical test to assess how well observed data fit expected results.

G/S checkpoint Control mechanism that prevents cells starting DNA replication with damaged DNA.

Guanine A purine nitrogenous base which pairs with cytosine.

Guanylyl transferase An enzyme that forms an unusual 5′ to 5′ phosphodiester bond between a guanine nucleotide and the 5′ terminal nucleotide during 5′ capping of pre-mRNA.

Hairpin loop An RNA secondary structure that forms when palindromic sequences base pair to form a short double helix stem that ends in an unpaired loop.

Glossary

Haploid A cell or an individual whose cells have a single set of chromosomes.

Haploinsufficiency The situation when only one functional allele is present and does not produce enough gene product for full function leading to an abnormal or diseased state.

Haplotype A set of closely linked alleles or single nucleotide polymorphisms that tend to be inherited together.

Hapmap project A project that aims to develop a detailed single nucleotide polymorphism haplotype map of the human genome to help gene mapping.

Hardy–Weinberg equilibrium The state in which genotype frequencies in a population are in accord with expectations of the Hardy–Weinberg law.

Hardy–Weinberg law The basic principle of population genetics which states that in a population of sexually reproducing, randomly mating diploid individuals, if a gene has two alleles of frequencies p and q, the frequencies of the homozygote genotypes will be p^2 and q^2 and that of the heterozygote will be $2pq$ within a single generation. These frequencies will be the same in future generations in the absence of mutation, migration, and selection.

Heat shock protein A class of proteins whose expression is increased when cells are exposed to elevated temperatures and other stresses. They assist new and distorted proteins to properly fold.

Helix-loop-helix A DNA binding domain of a transcription activator that consists of a shorter and longer alpha helix connected by a short irregular region.

Helix-turn-helix A DNA binding domain of a transcription activator that consists of two alpha helices connected by a short irregular region.

Hemizygous The presence of a single copy of a gene; generally refers to the single copy of the X chromosome in male mammals.

Heritability The proportion of the total variation in a trait that is due to genetic rather than environmental factors.

Heterochromatin Tightly coiled, genetically inactive chomatin.

Heterogametic sex The sex with two different sex chromosomes.

Heteroplasmy A cell or organism that has two or more alleles for a mitochondrial or chloroplast gene.

Heterozygous The state of having two different alleles of a gene.

Heterozygous advantage The situation in which the fitness of heterozygous individuals is greater than that of either type of homozygote.

Histones Low molecular weight basic proteins that DNA tightly coils around to form chromatin.

Histone code The idea that gene expression is in part regulated by the pattern of chemical modifications of histones.

Holliday junction The cross-like structure formed at a crossover between two DNA double helices.

Holoenzyme A complete functional multi-subunit enzyme. Often used in reference to RNA polymerase.

Homeobox A DNA sequence (~180 bp) within a gene used during development, which, when expressed, encodes a DNA binding domain.

Homogametic sex The sex with two identical sex chromosomes.

Homologue A chromosome that is similar in physical characters and genetic information to another chromosome with which it pairs during meiosis.

Homologous recombination The exchange of DNA segments between non-sister chromatids of homologous chromosomes at a crossover during prophase I of meiosis. It is also a mechanism for repairing damaged DNA.

Homoplasmy A cell or organism that has just one allele for a mitochondrial or chloroplast gene.

Homozygous The state of having two identical alleles of a gene.

Housekeeping gene A constitutive gene required for the maintenance of basic cellular function.

HU protein A low molecular weight histone-like protein that binds to and bends bacterial DNA, helping to package it in the nucleoid region of the cell.

Hybridization The process of forming a double-stranded nucleic acid molecule from two single strands of DNA and/or RNA. It also refers to the mating of two related species.

Ideogram A drawing of a chromosome showing its physical characters and banding pattern.

Imprinting The process by which maternally and paternally derived chromosomes are differently modified chemically. This results in the different expression of a certain gene(s) on these chromosomes depending on their parental origin.

Inbreeding Reproduction between related individuals.

Inbreeding coefficient (F) A measure of the level of inbreeding among individuals in a population.

Inbreeding depression The reduction in fitness of offspring from inbreeding. It results from the increased likelihood of the same deleterious recessive allele being inherited from both parents and thus being expressed in an individual.

Incomplete dominance The situation when heterozygotes show a phenotype intermediate between the phenotype of each homozygote.

Independent assortment The random distribution of each member of a pair of homologous chromosomes to gametes during meiosis. The distribution of one pair of alleles is independent of that of allele pairs on other homologous chromosomes.

Induced mutation A mutation produced by interaction of DNA with environmentally derived chemicals or radiation.

Inducer An effector molecule that activates transcription; generally refers to activation of a bacterial operon.

Inducible operon An operon that encodes proteins only required occasionally. Transcription is inhibited until events 'induce' removal of a repressor.

Initiation codon The mRNA triplet AUG that signals the start of translation and codes for the insertion of methionine (formylmethionine in bacteria) as the first amino acid in a polypeptide.

Initiation factor A protein required for the start of translation.

Insertion The gain of genetic material. The size of the gain varies from a single nucleotide to a segment of a chromosome.

Insertion sequence (IS) A short DNA sequence that acts as a simple transposon.

Integrase A viral-encoded enzyme that enables viral DNA to be integrated into the host cell chromosome.

Intercalating reagent A chemical that inserts between base pairs in a DNA molecule, disrupting complementary base pairing.

Interference The physical constraint to the formation of a crossover caused by a nearby crossover.

Internally transcribed spacers (ITSs) Non-functional RNA between 18S, 5.8S, and 28S rRNA that is removed during processing of the rRNA transcript.

Interphase The stage of the cell cycle between cell divisions. During interphase gene expression and DNA replication occur.

Glossary

Intron A non-coding sequence that interrupts the coding sequence of a gene.

Inversion The situation in which a segment of a chromosome is reversed.

Inversion heterozygote An individual in which one member of a pair of homologous chromosomes contains an inversion.

Inversion loop The looped structure formed during prophase I of meiosis by a pair of homologous chromosomes when one of the chromosomes contains an inversion.

Inverted repeats A DNA sequence that is identical to another sequence on the same chromosome except that it is oriented in the opposite direction.

Isoaccepting tRNAs Different tRNAs carrying the same amino acid.

Isochromosome A structural chromosome mutation in which one chromosome arm is lost and the other duplicated.

Isoform A different form of the same protein, generally produced by alternative splicing.

Jumping gene *see* Transposon.

Junk DNA Genomic DNA that does not code for a protein, RNA molecule or their regulation.

Karyogamy The fusion of nuclei during fertilization.

Karyotype A visual display of all the chromosome from a nucleus, arranged in pairs in descending size order.

Kinetochore A specialized region within the centromere to which spindle fibres attach.

Knock-in mouse A transgenic mouse in which expression of the inserted gene is controlled by RNA interference. There is often residual activity of the encoded protein.

Knockout mouse A transgenic mouse in which the inserted gene has a mutation which abolishes the function of the encoded protein.

Kozak sequence A nucleotide sequence (5'ACCAUGG3') in the 5' untranslated region of eukaryotic mRNA that enables ribosomes to recognize the initiator codon.

Lagging strand The DNA template strand that is replicated discontinuously.

Lampbrush chromosome A giant chromosome found in oocyte nuclei during prophase I of meiosis. Its brush-like appearance results from de-condensation of multiple transcribing genes.

Lariat The pre-mRNA is cut at the 5' splice donor site. This frees the 5' end of the intron, which attaches to the branch point: guanine nucleoside from the 5' end bonds with the free 2' OH of the adenine nucleoside at the branch point.

Law of Independent Assortment Mendel's second law, which states that different pairs of alleles assort independently of each other at gamete formation.

Law of Segregation Mendel's first law, which states that the two members of an allele pair separate from each other into different gametes.

Leader sequence An N-terminal signal sequence that directs secretion and processing of protein.

Leading strand The template strand that is replicated continuously.

Lethal allele An allele that eliminates a function essential for survival.

Leucine zipper A leucine-rich interacting domain of a family of transcription factors.

LINE (long interspersed element) A DNA sequence that is several thousand nucleotides long and repeated many times in the human genome.

Linkage Traits that are inherited together as their genes are situated close to one another on the same chromosome.

Linkage analysis A study that investigates whether two or more genes are linked.

Linker DNA The segment of DNA between two nucleosomes.

Locus (plural: loci) The physical location of a gene on a chromosome.

Locus heterogeneity The situation in which mutations in different genes cause the same phenotype.

LOD score (logarithm of odds) This is a statistical test that assesses the likelihood of linkage between two genes.

Lyonization The process by which one of the two X chromosomes in a female mammalian cell is inactivated.

Lysogenic cycle Prior to a lytic cycle, a phage's genome is incorporated into and replicated along with the bacterial chromosome for a number of generations.

Lytic cycle A bacteriophage reproductive cycle during which the host bacterial cell is killed during the production of several hundred new viruses.

Maternal imprinting A situation in which the maternal derived allele is silenced and paternal allele expressed.

Maternal influenced trait An individual's trait whose phenotype is determined by his/her mother's genotype.

Maternal inheritance A non-Mendelian pattern of inheritance in which all progeny have the genotype and phenotype of the female parent.

Mean The average value of a set of measurements obtained by summing all values and dividing this total by the number of measurements in the sample.

Mediator complex A multi-protein transcriptional coordinator.

Meiosis A type of cell division that results in four daughter cells each with half the number of chromosomes as the parental cell.

Melting temperature (T_M) The temperature at which half a DNA molecule exists in double-stranded form and half as separated, single strands.

Messenger RNA (mRNA) The form of RNA that carries the genetic instructions for synthesis of a protein.

Metacentric chromosome A chromosome with its centromere positioned centrally.

Metaphase A stage in mitosis and meiosis during which chromosomes are aligned at the equator in the middle of a cell.

Methylation The addition of a methyl group, e.g. to a nitrogenous base in DNA.

Microarray Grid of DNA fragments attached to a solid surface, which function as probes to detect the presence of complementary sequences.

Microdeletion Very small chromosomal deletions (<2 Mb).

MicroRNA (miRNA) Small RNA molecules (21–24 bp) that regulate translation of mRNA by binding reversibly to the 3′ untranslated region.

Microsatellite A short DNA sequence, usually 2–4 nucleotides, that is tandemly repeated. The number of repeats varies between individuals.

Migration The movement of organisms from one habitat to another.

Minisatellite A DNA segment that consists of tandem repeats of a sequence 10–100 nucleotides in length.

Mismatch repair A DNA repair system that detects and repairs incorrectly paired nucleotides.

Missense mutation A single nucleotide change that alters a codon and its encoded amino acid.

Glossary

Mitosis A type of cell division that results in two daughter cells each with the same chromosome number and genetic information as the parental cell.

Monohybrid cross A breeding experiment between two individuals that differ in the phenotypic expression of a single trait. The inheritance pattern of a pair of alleles is analysed.

Monosomy A diploid cell or organism lacking one chromosome $(2n-1)$.

Monozygous twin A twin derived from the splitting of a single zygote.

Mosaic The presence in one individual of tissues whose cells have different numbers of chromosomes.

Multifactorial A trait whose phenotype is determined by multiple genes and various environmental factors.

Multiple alleles Three or more possible alleles for a gene.

Multiple cloning site A short segment of a cloning vector that contains many restriction sites for insertion of DNA being cloned.

Multiplication rule The probability of the co-occurrence of two (or more) independent events is obtained by multiplying the probabilities of each independent event.

Mutagen An environmental agent (chemical or radiation) that increases the rate of mutation above the spontaneous rate.

Mutation A change in the genetic material.

Mutation hotspot A DNA sequence with a high susceptibility to mutation.

Narrow sense heritability The proportion of phenotypic variance that is due to additive genetic variance.

Natural selection The evolutionary process by which the better adapted organisms survive and reproduce.

Neutral mutation A mutation that has no effect upon the fitness of an individual.

Next generation sequencing New twenty-first century methods of DNA sequencing that achieve high throughput and can simultaneously parallel sequence 10^3–10^6 DNA fragments.

Nitrogenous base One of the three chemical components of a nucleotide.

Non-disjunction A failure of homologous chromosomes to separate at anaphase I of meiosis, or sister chromatids to separate during anaphase II of meiosis or during mitosis.

Nonsense mutation A single nucleotide change that alters a codon to a stop codon, truncating protein synthesis.

Nucleoside A nitrogenous base linked to a pentose sugar.

Nucleosome The basic structural unit of chromatin consisting of ~146 nucleotides of DNA wrapped around a core of 8 histone proteins.

Nucleotide The structural unit of DNA and RNA consisting of a nitrogenous base, pentose sugar, and phosphate.

Nucleotide excision repair A DNA repair system that detects and repairs damaged DNA that contains physical distortions, such as ultraviolet-induced pyrimidine dimers.

Nullisomy The loss of both copies of a homologous pair of chromosomes from a diploid complement.

Okazaki fragments The multiple sections of newly synthesized DNA on the lagging template strand.

Open reading frame A DNA sequence that contains a start and stop codon in the same reading frame, and which potentially encodes a protein.

Operator A segment of DNA downstream of an operon's promoter to which a repressor reversibly binds, switching the operon on and off.

Operon A functional group of bacterial genes whose transcription is regulated by the presence or absence of repressors.

Origin of replication A specific sequence in a chromosome at which replication is initiated.

Outbreeding The mating of unrelated individuals.

P arm The upper arm of a chromosome.

P bodies Enzyme associations in eukaryotic cytoplasm that regulate mRNA decay and RNA-mediated gene silencing.

P element A transposon present in the genome of the fruit fly *Drosophila melanogaster* that is used to produce transgenic flies.

Palindrome A DNA or RNA sequence that is repeated nearby in reverse complementary orientation.

Paracentric inversion A chromosomal inversion that does not include the centromere.

Paternal imprinting A situation in which the paternal derived allele is silenced and maternal allele expressed.

Pedigree A family tree showing the relationships between members of a family and which individuals express a particular trait.

Penetrance The proportion of individuals carrying a particular allele who express the associated phenotype.

Peptidyl (P) site The site on a ribosome where a peptide bond is formed.

Pericentric inversion A chromosomal inversion that includes the centromere.

Phage *see* Bacteriophage.

Phenotype The observable features of an organism resulting from the interaction between the organism's genotype and its environment.

Phenotypic variance The variation in phenotypic expression of a trait

Philadelphia chromosome The derived chromosome 22, resulting from a reciprocal translocation between chromosomes 9 and 22, which is characteristic of individuals with chronic myeloid leukaemia.

Phosphodiester bond The covalent bond that links nucleotides together. It forms between the 5′ phosphate group on one nucleotide and the 3′ hydroxyl group on an adjacent nucleotide.

Photolyase An enzyme that directly repairs pyrimidine dimers by breaking the cyclobutane bond that links the two pyrimidines.

Photoreactivation The process by which pyrimidine dimers are repaired by photolyase.

Phylogeny An evolutionary tree showing the relatedness of different species or genes.

Physical mapping The use of molecular analytic tools to precisely determine a gene's chromosomal locus.

Pilus A hollow hair-like protein bridge between the cytoplasm of two bacteria, through which an F plasmid moves from one bacterium to another.

Piwi RNA (piRNA) Forms RNA–protein complexes that silence retrotransposons in germ line cells.

Plasmid Small, circular, self-replicating DNA molecules present in single or multiple copies in the cytoplasm of bacterial cells.

Plasmogeny A stage in fungal sexual reproduction when cells fuse, but nuclei remain separate.

Glossary

Pleiotropy The situation in which one gene influences two or more traits.

Pole Either end of the main axis of a cell during mitosis or meiosis.

Poly(A) tail The sequence of 50–250 adenine nucleotides added to pre-mRNA during post-transcriptional processing.

Polycistronic A section of DNA or mRNA that codes for two or more polypeptides.

Polygenic A trait whose phenotypic expression is influenced by many genes.

Polymerase chain reaction (PCR) A method of amplifying specific DNA sequences by means of repeated rounds of primer-directed DNA synthesis.

Polymorphism The existence of two or more alleles at a locus: the frequency of the most common being 0.99 or less.

Polyploid A cell or organism with three or more sets of chromosomes.

Polysome A cluster of ribosomes bound to a single mRNA all actively engaged in protein synthesis.

Polytene Giant interphase chromosomes formed from repeated rounds of DNA replication without cell division.

Population A group of individuals of a species which are living and breeding together.

Positional cloning Following linkage analysis, the use of DNA sequencing and other techniques to locate a gene's chromosomal position.

Pre-initiation complex The association of the small ribosomal subunit, initiation factors and initiator tRNA during initiation of protein synthesis.

Pribnow box The –10 consensus sequence (TATAAT) recognized and bound by RNA polymerase during transcription initiation in prokaryotes.

Primase An enzyme that synthesizes a short RNA sequence that primes DNA synthesis during DNA replication

Primer A short DNA or RNA sequence that serves as a starting point for DNA synthesis.

Proband An individual who presents with a medical problem and who is the stimulus for construction of a pedigree.

Prokaryote A unicellular organism lacking membrane-bound organelles.

Promoter A specific DNA sequence upstream (5′) of a gene which signals to the RNA polymerase the initiation site for transcription.

Proofreading Exonuclease activity of a DNA polymerase that removes incorrectly incorporated nucleotides during DNA synthesis.

Prophage The DNA of a bacteriophage that has integrated into a bacterial chromosome.

Prophase The first stage in mitosis and meiosis when the nuclear envelope breaks down and chromosomes condense and become visible. In meiosis chromosome pairing and exchange occurs.

Proteome All the proteins produces by a species.

Prototroph A wild type bacteria able to grow on un-supplemented minimal media.

Pseudoautosomal regions Two small regions of homology between X and Y chromosomes. These pair during prophase I of meiosis ensuring X and Y chromosomes segregate into separate gametes.

Pseudodominance The situation where the inheritance of an autosomal recessive trait mimics an autosomal dominant pattern, generally due to deletion of the dominant allele.

Pseudogene A non-functional gene with sequence homology to known coding genes.

Punnett square A grid to represent the progeny formed in a genetic cross.

Pure breeding A group of individuals that produce progeny of the same phenotype when selfed or interbred.

Purine A double-ringed nitrogenous base found in DNA and RNA.

Pyrimidine A single-ringed nitrogenous base found in DNA and RNA.

Pyrimidine dimers Ultraviolet-induced bonding of adjacent pyrimidine molecules in a strand of DNA.

Pyrophosphate (PP$_i$) Two phosphate groups produced during formation of a phosphodiester bond.

Pyrosequencing A recently developed DNA sequencing technique that detects the pyrophosphate as it is produced during DNA synthesis.

Q arm The lower arm of a chromosome.

Quantitative trait A trait whose phenotypic expression shows a wide range of expression between two extremes, and is influenced by many genes and environmental factors.

Quantitative trait locus (QTL) Gene or chromosomal locus that contributes to expression of a quantitative trait.

R plasmid A bacterial plasmid that carries genes that confer resistance to various antibiotics.

Reading frame The particular way a nucleotide sequence is read during translation. Nucleotides are decoded in groups of three (codons), beginning with a start codon and ending with a stop codon.

Real time PCR A PCR technique that quantifies the target DNA as it is amplified.

Recessive An allele that is only expressed phenotypically when its paired allele is identical.

Recessive epistasis The situation when expression of alleles at one locus is suppressed in individuals which are recessive homozygotes at a second, epistatic locus.

Reciprocal crosses A pair of crosses in which the phenotypes of the male and female parents are reversed.

Reciprocal translocation Reciprocal exchange of DNA segments between two non-homologous chromosomes.

Recombinant DNA technology The transfer of a DNA fragment from one organism to a self-replicating genetic element that replicates the fragment in a foreign host cell.

Recombination A process, generally involving crossing over during meiosis, that results in new allele combinations.

Repetitive DNA DNA sequences that are present in multiple copies in a genome.

Replica plating A technique for the transfer of colonies of bacteria or fungi from one agar Petri plate to another, such that their relative positions are maintained.

Replication bubble A section of a double helix that is unwound and undergoing DNA replication.

Replication fork The Y-shaped point at which a DNA double helix separates into two single strands, each of which forms a template, during DNA replication.

Replication licensing system A complex of regulatory proteins that ensure DNA replication is only initiated once per cell cycle.

Repressible operon An operon that encodes proteins constitutively required. Transcription occurs unless a repressor binds to the operator.

Repressor A regulatory molecule that inhibits transcription; generally refers to inhibition of a bacterial operon.

Response to selection The amount by which a quantitative trait changes its phenotypic expression in a single generation of selective breeding.

Restriction enzyme An enzyme that recognizes specific DNA sequences and makes double-stranded cuts.

Restriction site The specific DNA sequence at which a restriction enzyme cleaves DNA

Glossary

Restriction site fragment polymorphism The variable-sized fragments produced in different individuals when DNA is cut with the same restriction enzyme.

Retrotransposon Class I transposons that produce DNA copies for insertion in new locations in the genome via transcription and reverse transcription.

Retrovirus An RNA virus that replicates its genome via a DNA copy, which often integrates into the host cell genome.

Reverse genetics One approach to identifying a gene's function: starting with a particular genotype (or DNA sequence) the associated phenotype is deduced.

Reverse transcriptase An enzyme which synthesizes complementary DNA (cDNA) from an RNA template.

Rho factor An RNA-dependent ATPase that promotes termination of prokaryotic transcription by destabilizing the interaction between template DNA and nascent mRNA, thus promoting release of the newly synthesized mRNA.

Ribonuclease An enzyme that hydrolyses RNA.

Ribonucleoprotein A functional complex of RNA and protein.

Ribose A pentose sugar which is a constituent of ribonucleotides.

Ribosomal RNA (rRNA) The RNA component of ribosomes.

Ribosome An organelle made of RNA and proteins that assembles amino acids into polypeptides.

Riboswitch A regulatory sequence in an RNA molecule that assumes different secondary structures in response to the binding of regulatory proteins.

Ribozyme An RNA molecule with a catalytic function.

Ring chromosome A circularized chromosome in which the two telomeres have joined.

RNA interference An RNA-based mechanism for controlling gene expression. Small double-stranded RNA molecules bind to and inhibit translation of mRNA.

RNA polymerase An enzyme that catalyses RNA synthesis from a DNA template.

Robertsonian translocation The fusion of two acrocentric chromosomes at their centromeres.

S phase The stage of the cell cycle during which DNA replication occurs.

Satellite DNA Large arrays of tandemly repeated non-coding DNA.

Selectable marker A gene carried by a vector and conferring a recognizable characteristic, e.g. antibiotic resistance, which enables a host cell containing the vector to be identified.

Selection coefficient A measure, used in population genetics, of the relative fitness of different genotypes.

Selection differential The difference between the average phenotype of individuals selected for a breeding programme and the average population phenotype.

Self-splicing An intron forms a ribozyme that catalyses its own excision from the RNA transcript.

Sense strand This is the complementary DNA strand to the template strand for transcription of a gene. It has the same sequence as the transcribed mRNA.

Sex chromosome A chromosome that determines the sex of an individual.

Sex-influenced trait An autosomal trait that is expressed differently in males and females.

Sex-limited trait A trait expressed in only one sex.

Sex-linked trait A trait encoded by a gene on a sex chromosome that shows different patterns of expression in males and females.

Shine–Dalgarno sequence A consensus sequence (AGGAGG), in the 5′ untranslated region of prokaryotic mRNA, identifying the ribosome binding site.

Sigma factor A mobile subunit of prokaryotic RNA polymerase that recognizes and binds the enzyme to the promoter for initiation of transcription.

Signal sequence An N-terminal sequence of amino acids that directs the processing or cellular localization of a protein.

SINE (short interspersed element) A DNA sequence that is several hundred nucleotides long and repeated many times in the human genome.

Single nucleotide polymorphism (SNP) A single nucleotide variant in a DNA sequence.

Sister chromatids The two identical chromatids of a replicated chromosome, joined by a centromere.

Site-specific mutagenesis A technique to create a mutation at a defined site in a DNA sequence.

Small nuclear RNA (snRNA) A form of RNA complexed with proteins to form spliceosomes.

Small nucleolar RNA (snoRNA) RNA which processes and chemically modifies rRNA.

Solenoid A transcriptionally inactive form of chromatin in which nucleosomes are coiled and compacted to form a fibre of 30 nm diameter.

SOS response An error-prone DNA repair system triggered by gross DNA damage.

Spindle A network of microtubules that move and segregate pairs of chromatids and chromosomes during mitosis and meiosis.

Splice acceptor site (3′ splice site) A consensus sequence at the 3′ end of an intron that is recognized and cut by a spliceosome during intron splicing.

Splice donor site (5′ splice site) A consensus sequence at the 5′ end of an intron that is recognized and cut by a spliceosome during intron splicing.

Spliceosome A complex of five snRNAs and many proteins that excises introns from pre-mRNAs.

Spontaneous mutation Mutations that occur as the result of natural processes in the cell.

Sporophyte In plants, the diploid spore-producing generation.

SRY The male sex-determining gene on the Y chromosome.

Standard deviation A statistical index that describes the variability within a normal distribution. It predicts the percentage of samples which fall within certain ranges relative to the mean.

Start codon The codon (AUG) on mRNA where translation begins.

Stem loop *see* Hairpin loop.

Stop codon Three codons (UAA/UAG/UGA) on mRNA that signal termination of translation.

STRs (short tandem repeats) *see* Microsatellite.

Supercoiled DNA A twisted tertiary DNA structure that forms when the double helix is under- or over-wound.

Synapsis The close pairing of homologous chromosomes during prophase I of meiosis.

T$_i$ plasmid Plasmid of *Agrobacterium tumifaciens*. A useful cloning vector for introducing foreign DNA into plant genomes.

TATA box *see* Pribnow box.

Tandem repeat A segment of DNA in which two or more nucleotides are multiply repeated and the repetitions lie adjacent to each other.

Glossary

Tautomerization A spontaneous change in proton position in a nitrogenous base that produces a structural isomer with different base-pairing properties.

Telocentric A chromosome with its centromere close to one end.

Telomerase An enzyme that catalyses replication of a telomere.

Telomere Repetitive non-coding DNA situated at the ends of eukaryotic chromosomes, protecting them from degradation.

Temperate phage A bacteriophage genome that has integrated into, and is replicated at the same time as, the host bacterial chromosome.

Template A DNA or RNA strand whose sequence is copied into a new complementary strand.

Test cross A cross between an individual of dominant phenotype, but unknown genotype, with another who is homozygous recessive.

Tetravalent A structure formed during prophase I of meiosis when two pairs of homologous chromosomes, variously sharing homology, pair.

Thermocycler A machine used to amplify segments of DNA via the polymerase chain reaction.

Threshold trait A trait which is inherited quantitatively, but which shows qualitative expression; generally, two alternative phenotypes.

Thymine A pyrimidine nitrogenous base which pairs with adenine.

Thymine dimer A DNA lesion caused by a pair of adjacent covalently linked thymines produced by exposure to ultraviolet light.

Topoisomerase An enzyme that regulates the level of supercoiling in DNA.

Transcription The process of synthesizing RNA from a DNA template.

Transcription bubble A short DNA segment that has unwound to expose single-stranded DNA for transcription.

Transcription-coupled NER A form of nucleotide excision repair that preferentially targets DNA being actively transcribed.

Transcription factor A protein that binds to specific DNA sequences in regulatory regions helping to control gene expression.

Transcriptome All the RNA molecules transcribed from a genome.

Transduction The transfer of genes between bacteria by bacteriophages.

Transfection The mechanism by which DNA is introduced into mammalian cells from the external medium.

Transfer RNA (tRNA) The form of RNA that carries amino acids to the ribosomes and transfers them to the growing polypeptide chain during translation.

Transformation The uptake of DNA by a bacterium from the surrounding medium.

Transgenic An organism whose genome contains one or more foreign genes introduced by recombinant DNA techniques.

Transition A nucleotide substitution which results in a purine base being replaced by a different purine, or a pyrimidine by a different pyrimidine.

Translation The process by which amino acids are assembled at a ribosome to form a polypepetide.

Translocation The movement of a chromosomal segment to a non-homologous chromosome. Also refers to the movement of a ribosome along mRNA during translation.

Transposable element A DNA segment that is capable of self-generated movement from one site in the genome to another location.

Transposition Movement of a transposable element within a genome.

Transversion A nucleotide substitution which results in a purine base being replaced by a pyrimidine, or a pyrimidine by a purine.

Trihybrid cross A genetic cross that simultaneously analyses the inheritance patterns of alleles of three genes.

Trinucleotide repeat Tandem repeats of three nucleotides. Expanded copy numbers within genes result in inherited diseases.

Trisomic rescue The early embryonic loss of an extra chromosome from the cells of a trisomic embryo.

Trisomy The presence of an extra copy of a chromosome.

Trivalent The structure formed during prophase I of meiosis when three homologous copies of a chromosome pair.

tRNA charging The chemical reaction in which an aminoacyl-tRNA synthetase attaches an amino acid to its corresponding tRNA.

True breeding *see* Pure breeding.

Tubulin Protein constituent of microtubules.

Ubiquitin A eukaryotic regulatory protein that tags defective proteins for degradation.

Unbalanced gamete Gamete with missing or extra chromosomes or chromosomal segments.

Unequal crossing over Misalignment of two DNA molecules during crossing over resulting in one DNA molecule with an insertion and the other with a deletion.

Uniparental disomy (UPD) The situation in which both members of an allele pair are inherited from one parent.

Uracil A pyrimidine nitrogenous base found in RNA and which pairs with adenine.

Uracil glycosylase A DNA repair enzyme that recognizes and removes a misincorporated uracil.

VNTRs (variable number of tandem repeats) *see* Microsatellites.

Variance A statistical index that describes the variability of a set of measurements.

Vector A small, replicating DNA molecule into which a foreign DNA fragment can be inserted and transferred to a host cell.

Wild type The phenotype or allele that is most commonly found in a natural population. It generally indicates the fully functioning, non-mutant variant of a trait.

Wobble The flexibility in base pairing between nucleotide 3 of an mRNA codon and nucleotide 1 of a tRNA anticodon, resulting in some tRNAs recognizing more than one mRNA codon.

X chromosome A sex-determining chromosome.

X-linked trait A trait determined by a gene located on the X chromosome.

Y chromosome A sex-determining chromosome.

Y-linkage A trait determined by a gene located on the Y chromosome.

Yeast artificial chromosome (YAC) A cloning vector that includes a yeast centromere, a pair of telomeres, and an origin of replication, and which carries large DNA segments (up to 3 Mb).

Z-DNA A form of DNA in which the double helix winds in a left-handed direction.

Zinc finger A DNA binding domain that includes one or more zinc ions.

Zoo blot Hybridization between DNA of related species. It is used to identify the function of an unknown coding sequence.

INDEX

Down syndrome (trisomy 21)
229, 244, 245
Drosophila melanogaster (fruit fly)
alternative splicing of *Dscam*
202
developmental gene analysis
294
dihybrid crosses 29–30, 32
gene duplication causing
Bar phenotype 239
linked genes 64, 65
sex-linked inheritance of eye
colour 43–4
drug resistance in bacteria
105–6, 107
Dscam, alternative splicing 202
duplicating genes 50
duplication
of chromosomes 237, 238–9
of genes 133, 239

E

Edward syndrome (trisomy 18)
229
elongation
DNA replication 143–5, 148
transcription 159, 164
translation 182–4
elongation factors 182
phosphorylation 206
embryonic stem (ES) cells
294–5
endoplasmic reticulum (ER) 176
endosymbiotic theory 120–1
enhancer sequences 161, 198
environmental factors in twin
studies 97–8
environmental variance (V_E)
87, 89
epigenetics 11, 128, 201
episomes 104
epistasis 47–51
Escherichia coli
genome size 101, 132
hypermutable strains 223
lac and *trp* operons 193–6
Eubacteria 2
see also prokaryotes
euchromatin 6
eukaryotes 2, 102
cell cycle 12–21, 227
chromosomes 2–11

control of gene expression
197–208
DNA replication 142, 146,
148, 149, 223
genome size 132
ribosomes 175, 176
RNA processing 165–70
sex determination 8–12
transcription 156, 157,
160–70, 198–202
transfection 113
translation 179–80, 182–5
eukaryotic viruses 117–20
evolution
gene duplication 133, 239
natural selection 268–70
phylogenetics 158, 177, 292
polyploidy 233
population genetics 258–60,
261, 266, 268–70, 272
exon skipping 202
exosomes 202
exportins 203
expressed sequence tags (ESTs)
280
expression vectors 276
expressivity 54

F

F plasmids 104–5, 109–10
F1 generation 23, 29
F2 generation 23, 29, 38, 43, 47
females
sex determination 8–9,
11–12
sex-linked traits 11, 42–5, 53
threshold trait variation 95
fitness coefficient (w) 269
5′ cap of mRNA 167, 180
fluorescent in situ hybridization
(FISH) 249–50
forensic investigations 134–5,
215
forward genetics 293–4
founder effect 272
four o'clock plant (*Mirabilis
japonica*), maternal
inheritance of leaf colour
45, 123
fragile sites 248
frameshift mutations 211–12,
216

free radicals 216, 219
fruit fly *see Drosophila
melanogaster*
fungi
life cycle 21
maternal inheritance 45

G

G-banding 6
G_0 phase 13
G_1 phase 13, 149
G_2 phase 13
Gal 4 transcription activator
199–200
gametes 3
meiosis 16–20
recombinant 60–2, 67–8
unbalanced 230, 234,
239, 241
gametophytes 22
gel electrophoresis 286–7
geminin 149
gene dosage imbalance 229,
238–9
gene expression, control 188–9
eukaryotic 197–208
post-transcriptional 202–4
post-translational 206–8
prokaryotic 192–7
regulation points 189
transcriptional 190–6,
198–202
translational 192, 196–7,
204–6
gene families 133
gene function, deducing 293–8
gene mapping 59
in bacteria 109, 110–12,
113, 117
dihybrid crosses 62–9
DNA markers 75–7, 99–100
linkage of genes 60–1
LOD score (pedigree analysis)
72–5
physical mapping 77–8
trihybrid crosses 69–72
gene transfer in bacteria
108–13, 115–17
genetic code 172–3
genetic conflict hypothesis 58
genetic cross diagrams 25–6,
42–3

prophages 115, 117
prophase
  meiosis 17, 19, 214
  mitosis 16
proteins 173–4
  post-translational modification 186–7, 206–8
  synthesis *see* translation
proteome 207–8
proviruses 119–20
pseudoautosomal region of sex chromosomes 10
pseudodominance 238–9
pseudogenes 133
Punnett square 26, 43, 44
purines 127, 154
  depurination 214, 215
  methylation 128, 223
  tautomerization 212
pyrimidines 127, 153
  dimers 218, 219–20
  modifications 128, 214–15
  tautomerization 212
pyrosequencing 290–1

Q

q arm 3, 245
QTLs (quantitative trait loci) 98–100
quantitative traits 79–80
  additive genes 81–2
  heritability 87–94, 96–8
  human 93–100
  statistics 82–6

R

R plasmids 105–6
r-value (correlation coefficient) 84–5
radiation, as a mutagen 218
random amplification of polymorphic DNA (RAPD-PCR) 284
reading frame 172, 211
real-time PCR (RT-PCR) 285
recessive alleles 24, 53
recessive epistasis 47–8
reciprocal crosses 23, 43–4, 45
reciprocal translocation 242, 243–4

recombinant DNA technology (cloning) 106, 274–81, 295–6
recombination 17, 60–1
  in bacteria 108, 110
  homologous, in DNA repair 224–6
  see also crossover, in eukaryotes
regression 85–6, 91
regulatory promoters 161
regulons 196
release factors 184
repetitive DNA 75–7, 134–5, 213
replica plating 106–7
replication of DNA 13, 122, 140–50
  errors 212–18
  PCR 281–6
  signal that DNA is newly synthesized 223
repressible operons 192, 195–6
repressors 192
reptiles, sex determination 11–12
repulsion (trans) configuration 65
response to selection (R) 91–3
restriction enzymes 114, 275–6, 277
restriction fragment length polymorphisms 76
retinoblastoma 54
retrotransposons 137
retroviruses 118–20, 137
reverse genetics 294–8
reverse transcriptase 119, 137
rho factor 160
ribose 126, 153
ribosomal RNA *see* rRNA
ribosomes 175–7
  translation 179–85
riboswitches 196
ribozymes 156, 170
ring chromosomes 244–5
RNA
  antisense 196–7
  central dogma 151–2
  genetic code 172–3
  miRNA 204–5
  priming of DNA replication 143, 144

snRNA 168
  structure 126, 130, 152–6
  types 153
  see also mRNA; rRNA; tRNA
RNA interference (RNAi) 204–5, 298
RNA polymerases 156–7, 158, 159, 160, 163, 164
RNA viruses 118–20
RNA-induced protein complexes (RISCs) 205
Robertsonian translocation 242–3, 244
rolling circle DNA replication 145
rRNA (ribosomal RNA) 175
  genes 134, 176
  processing 176–7
RT-PCR (real-time PCR) 285

S

S phase 13, 149
Sanger method for DNA sequencing 287–9
satellite DNA 75–7, 134–5, 213
segmented genomes 117
segregation 27, 31
  after chromosomal translocations 243–4
selection
  artificial (selective breeding) 91–3
  natural 268–70
selection coefficient (s) 269
selection differential (S) 91
self-splicing 170
semi-conservative replication of DNA 140–1
sequencing of DNA 287–92
sex determination
  genetic 8–11
  non-genetic 11–12
sex-influenced traits 44–5
sex-limited traits 44
sex-linked traits 42–4, 53
  allele frequency calculations 257
Shine–Dalgarno sequence 166, 181
short interfering RNA (siRNA) 204–5